Rudolf...

The Continuum

Rudolf Taschner

The Continuum

A Constructive Approach
to Basic Concepts of Real Analysis

Bibliographic information published by Die Deutsche Bibliothek
Die Deutsche Bibliothek lists this publication in the Deutsche Nationalbibliographie;
detailed bibliographic data is available in the Internet at <http://dnb.ddb.de>.

Prof. Dr. Rudolf Taschner
Institute for Analysis and
Scientific Computing
Vienna University of Technology
Wiedner Hauptstr. 8
A-1040 WIEN

E-Mail: rudolf.taschner@tuwien.ac.at

First edition, September 2005

All rights reserved
© Friedr. Vieweg & Sohn Verlag/GWV Fachverlage GmbH, Wiesbaden 2005
Softcover reprint of the hardcover 1st edition 2005

Editorial office: Ulrike Schmickler-Hirzebruch / Petra Rußkamp

Vieweg is a company in the specialist publishing group Springer Science+Business Media.
www.vieweg.de

No part of this publication may be reproduced, stored in a retrieval system
or transmitted, mechanical, photocopying or otherwise without prior
permission of the copyright holder.

Cover design: Ulrike Weigel, www.CorporateDesignGroup.de
Printed on acid-free paper
ISBN-13: 978-3-322-82038-9   e-ISBN-13: 978-3-322-82036-5
DOI: 10.1007/978-3-322-82036-5

Preface

"Few mathematical structures have undergone as many revisions or have been presented in as many guises as the real numbers. Every generation re-examines the reals in the light of its values and mathematical objectives." This citation is said to be due to Gian-Carlo Rota, and in this book its correctness again is affirmed. Here I propose to investigate the structure of the mathematical continuum by undertaking a rather unconventional access to the real numbers: the intuitionistic one. The traces can be tracked back at least to L.E.J. Brouwer and to H. Weyl. Largely unknown photographies of Weyl in Switzerland after World War II provided by Peter Bettschart enliven the abstract text full of subtle definitions and sophisticated estimations.

The book can be read by students who have undertaken the usual analysis courses and want to know more about the intrinsic details of the underlying concepts, and it can also be used by university teachers in lectures for advanced undergraduates and in seminaries for graduate students.

I wish to thank Walter Lummerding and Gottfried Oehl who helped me with their impressive expert knowledge of the English language. I also take the opportunity to express my gratitude to Ulrike Schmickler-Hirzebruch and to the staff of Vieweg-Verlag for editing my manuscript just now, exactly 50 years after the death of Hermann Weyl, in their renowned publishing house.

Vienna, 2005   *Rudolf Taschner*

Contents

1 Introduction and historical remarks   1
   1.1   FAREY fractions 1
   1.2   The pentagram 3
   1.3   Continued fractions 6
   1.4   Special square roots 8
   1.5   DEDEKIND cuts 9
   1.6   WEYL's alternative 12
   1.7   BROUWER's alternative 13
   1.8   Integration in traditional and in intuitionistic framework 15
   1.9   The wager 17
   1.10 How to read the following pages 19

2 Real numbers   21
   2.1   Definition of real numbers 21
       2.1.1   Decimal numbers 21
       2.1.2   Rounding of decimal numbers 23
       2.1.3   Definition and examples of real numbers 24
       2.1.4   Differences and absolute differences 26
   2.2   Order relations 27
       2.2.1   Definitions and criteria 27
       2.2.2   Properties of the order relations 29
       2.2.3   Order relations and differences 31
       2.2.4   Order relations and absolute differences 32
       2.2.5   Triangle inequalities 33

		2.2.6	Interpolation and Dichotomy	35
	2.3	Equality and apartness		38
		2.3.1	Definition and criteria	38
		2.3.2	Properties of equality and apartness	40
	2.4	Convergent sequences of real numbers		41
		2.4.1	The limit of convergent sequences	41
		2.4.2	Limit and order	42
		2.4.3	Limit and differences	44
		2.4.4	The convergence criterion	46

3 Metric spaces   49

	3.1	Metric spaces and complete metric spaces		49
		3.1.1	Definition of metric spaces	49
		3.1.2	Fundamental sequences	51
		3.1.3	Limit points	54
		3.1.4	Apartness and equality of limit points	57
		3.1.5	Sequences in metric spaces	58
		3.1.6	Complete metric spaces	60
		3.1.7	Rounded and sufficient approximations	61
	3.2	Compact metric spaces		64
		3.2.1	Bounded and totally bounded sequences	64
		3.2.2	Located sequences	65
		3.2.3	The infimum	67
		3.2.4	The hypothesis of DEDEKIND and CANTOR	70
		3.2.5	Bounded, totally bounded, and located sets	71
		3.2.6	Separable and compact spaces	72
		3.2.7	Bars	74
		3.2.8	Bars and compact spaces	76
	3.3	Topological concepts		78
		3.3.1	The cover of a set	78
		3.3.2	The distance between a point and a set	79
		3.3.3	The neighborhood of a point	80
		3.3.4	Dense and nowhere dense	82
		3.3.5	Connectedness	84
	3.4	The s-dimensional continuum		85
		3.4.1	Metrics in the s-dimensional space	85
		3.4.2	The completion of the s-dimensional space	86
		3.4.3	Cells, rays, and linear subspaces	89
		3.4.4	Totally bounded sets in the s-dimensional continuum	90
		3.4.5	The supremum and the infimum	90
		3.4.6	Compact intervals	92

4 Continuous functions   95

	4.1	Pointwise continuity		95
		4.1.1	The concept of function	95

	4.1.2	The continuity of a function at a point 96
	4.1.3	Three properties of continuity 98
	4.1.4	Continuity at inner points 102
4.2	Uniform continuity 105
	4.2.1	Pointwise and uniform continuity 105
	4.2.2	Uniform continuity and totally boundness 107
	4.2.3	Uniform continuity and connectedness 107
	4.2.4	Uniform continuity on compact spaces 109
4.3	Elementary calculations in the continuum 110
	4.3.1	Continuity of addition and multiplication 110
	4.3.2	Continuity of the absolute value 111
	4.3.3	Continuity of division 113
	4.3.4	Inverse functions 115
4.4	Sequences and sets of continuous functions 118
	4.4.1	Pointwise and uniform convergence 118
	4.4.2	Sequences of functions defined on compact spaces 121
	4.4.3	Spaces of functions defined on compact spaces 122
	4.4.4	Compact spaces of functions 124

5 Literature   **129**

Index   **134**

Hermann Hesse the Author of „The Glass Bead Game" and Hermann Weyl
(© Peter Bettschart, Wien)

Hermann Weyl (© Fr Schmelhaus, Zürich)

Hermann Weyl (© Peter Bettschart, Wien)

Hermann Weyl (© Peter Bettschart, Wien)

LEJ Brouwer (© E van Moorkorken 1943)

Hermann Weyl (© Peter Bettschart, Wien)

1
Introduction and historical remarks

It will come as no surprise to the reader to note that the title "The Continuum" refers to HERMANN WEYL's renounced book on the continuum, and in fact: the author of this book, though light-years away from the mathematical and philosophical capabilities of WEYL, shares his scepticism about the foundation of analysis in the sense of GEORG CANTOR and RICHARD DEDEKIND.

In order to understand the sceptical position of WEYL, it is advisable to call the idea of DEDEKIND cuts to our mind. We will do this by enumerating all rational numbers, and will show how unattainably sharp-edged DEDEKIND cuts are supposed to be.

1.1 FAREY fractions

In the beginning of the nineteenth century, the geologist JOHN FAREY constructed a table of fractions in the following way: In the first row he wrote $0/1$ and $1/1$. For $k = 2, 3, \ldots$ he used the rule: Form the k-th row by copying the $(k-1)$-st in order, but insert the fraction $(p+q)/(n+m)$, the so-called *mediant*, between the consecutive fractions p/n and q/m of the $(k-1)$-st row if $n + m \leq k$. Thus, since $1 + 1 \leq 2$ FAREY inserted $(0+1)/(1+1)$ between $0/1$ and $1/1$ and obtained $0/1, 1/2, 1/1$ for the second row. His third row was $0/1, 1/3, 1/2, 2/3, 1/1$. To obtain the fourth row he inserted $(0+1)/(1+3)$ and $(2+1)/(3+1)$ but not $(1+1)/(3+2)$ and $(1+2)/(2+3)$. The first five rows of FAREY's table were:

$$\frac{0}{1} \qquad\qquad\qquad\qquad\qquad\qquad\qquad \frac{1}{1}$$

$$\frac{0}{1} \qquad\qquad\qquad \frac{1}{2} \qquad\qquad\qquad \frac{1}{1}$$

$$\frac{0}{1} \qquad \frac{1}{3} \qquad \frac{1}{2} \qquad \frac{2}{3} \qquad \frac{1}{1}$$

$$\frac{0}{1} \quad \frac{1}{4} \quad \frac{1}{3} \quad \frac{1}{2} \quad \frac{2}{3} \quad \frac{3}{4} \quad \frac{1}{1}$$

$$\frac{0}{1} \ \frac{1}{5} \ \frac{1}{4} \ \frac{1}{3} \ \frac{2}{5} \ \frac{1}{2} \ \frac{3}{5} \ \frac{2}{3} \ \frac{3}{4} \ \frac{4}{5} \ \frac{1}{1}$$

Up to this row, at least, the table has a number of interesting properties. All the fractions that appear are in reduced form; all reduced fractions p/n between zero and one such that $n \leq k$ appear in the k-th row; if p/n and q/m are consecutive fractions in the k-th row, then $qn - pm = 1$ and $n + m > k$. We will prove all these properties for an even enlarged table: FAREY, of course, could have started with all integers ..., $-3, -2, -1, 0, 1, 2, 3, \ldots$ written as fractions ..., $-3/1$, $-2/1, -1/1, 0/1, 1/1, 2/1, 3/1, \ldots$ in the first row. We denote this first row by $\mathbb{Q}_1$. Then we construct, for any positive integer k, the k-th row from the $(k-1)$-st by the same rule as before, and we denote this k-th row by $\mathbb{Q}_k$.

If p/n and q/m are consecutive fractions in $\mathbb{Q}_k$, say with p/n to the left of q/m, then $qn - pm = 1$.

Proof. It is true for $k = 1$. Suppose it is true for the $(k-1)$-st row $\mathbb{Q}_{k-1}$. Any consecutive fractions in the k-th row $\mathbb{Q}_k$ will be either $p/n, q/m$ or p/n, $(p+q)/(n+m)$, or $(p+q)/(n+m), q/m$ where p/n and q/m are consecutive fractions in $\mathbb{Q}_{k-1}$. But then we have $qn - pm = 1$, $(p+q)n - p(n+m) = qn - pm = 1$, $q(n+m) - (p+q)m = qn - pm = 1$, and the assertion is proved by mathematical induction. ∎

Two immediate consequences are:

Every fraction in the table is in reduced form.

The fractions in each row are listed in order of their size.

Next we prove the following fact:

If p/n and q/m are consecutive fractions in $\mathbb{Q}_k$, then among all fractions with values between these two, the mediant $(p+q)/(n+m)$ is the unique fraction with the smallest denominator.

Proof. In the first place, the mediant will be the first fraction to be inserted between p/n and q/m as we continue to further rows of the table. It will appear in the $(n+m)$-th row. Therefore we have

$$\frac{p}{n} < \frac{p+q}{n+m} < \frac{q}{m}.$$

Now consider any fraction r/l between p/n and q/m so that $p/n < r/l < q/m$. Then

$$\begin{aligned}\frac{q}{m} - \frac{p}{n} &= \left(\frac{q}{m} - \frac{r}{l}\right) + \left(\frac{r}{l} - \frac{p}{n}\right) \\ &= \frac{ql-rm}{ml} + \frac{rn-pl}{nl} \geq \frac{1}{ml} + \frac{1}{nl} = \frac{n+m}{nml},\end{aligned}$$

and therefore

$$\frac{n+m}{nml} \leq \frac{qn-pm}{nm} = \frac{1}{nm},$$

which implies $l \geq n+m$. If $l > n+m$ then r/l does not have least denominator among fractions between p/n and q/m. If $l = n+m$, then the $\geq$-sign above reduces to the equality sign and we have $ql - rm = 1$ and $rn - pl = 1$. Solving, we find $r = p+q$, and hence $(p+q)/(n+m)$ is the unique fraction lying between p/n and q/m with denominator $n+m$. ∎

Any reduced fraction r/l with an integer r and a positive integer l appears in all $\mathbb{Q}_k$ with $k \geq l$.

Proof. This is obvious if $l = 1$. Suppose it is true for $l - 1$ with $l > 1$. Then the fraction r/l cannot be in the $(l-1)$-st row by definition and so it must lie in value between two consecutive fractions p/n and q/m of $\mathbb{Q}_{l-1}$. Thus $p/n < r/l < q/m$. Since

$$\frac{p}{n} < \frac{p+q}{n+m} < \frac{q}{m}$$

and $p/n, q/m$ are consecutive, the mediant $(p+q)/(n+m)$ is not in $\mathbb{Q}_{l-1}$ and hence $n+m > l-1$ by our induction hypothesis. The theorem above implies $l \geq n+m$, and so we have $l = n+m$. The uniqueness assertion of the theorem above shows that $r = p+q$. Therefore $r/l = (p+q)/(n+m)$ enters in $\mathbb{Q}_l$, and it is then in all later rows. ∎

1.2 The pentagram

Legend has it that PYTHAGORAS OF SAMOS himself taught that each ratio in the universe can be reduced to a fraction of integers, i.e. to a member of $\mathbb{Q}_k$ for a sufficient big number k. Legend also has it that HIPPASOS OF METAPONT, a member of the Pythagorean school, succeeded in refuting the hypothesis of PYTHAGORAS by observing the pentagram and calculating the ratio φ of the length d of its diagonal to the length s of its lateral: The pentagram arises from drawing the

4   1. Introduction and historical remarks

diagonals of a regular pentagon, and the points of intersection of these diagonals themselves define a new regular pentagon with s' as length of its laterals and with d' as length of its diagonals.

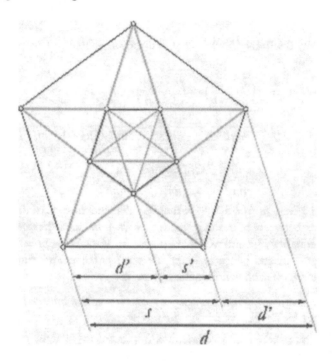

We have
$$\varphi = \frac{d}{s} = \frac{d'}{s'} \quad \text{and} \quad \frac{d}{d'} = \frac{s}{s'}$$
because both pentagrams are similar. Their symmetry further proves
$$d = s + d' \quad \text{and} \quad s = d' + s'.$$
Thus we have
$$\frac{d}{s} = \frac{d}{d - d'} = \frac{d - d' + d'}{d - d'} = 1 + \frac{d'}{d - d'} = 1 + \frac{1}{\frac{d}{d'} - 1}$$
$$= 1 + \frac{1}{\frac{s}{s'} - 1} = 1 + \frac{s'}{s - s'} = \frac{s - s' + s'}{s - s'} = \frac{s}{s - s'} = \frac{s}{d'}.$$

In other words: The four lengths d, s, d', s' that occur in the pentagram define three times the same ratio
$$\varphi = \frac{d}{s} = \frac{s}{d'} = \frac{d'}{s'}.$$

But the most remarkable relation is the following:
$$\varphi = \frac{d}{s} = \frac{s+d'}{s} = 1 + \frac{d'}{s} = 1 + \frac{1}{\frac{s}{d'}} = 1 + \frac{1}{\varphi}.$$

Suppose we start with the fraction $p_0/n_0 = 1/1$, i.e. $p_0 = 1, n_0 = 1$, and we define by induction for $k = 1, 2, 3, \ldots$

$$\frac{p_k}{n_k} = 1 + \frac{1}{\frac{p_{k-1}}{n_{k-1}}} = \frac{p_{k-1} + n_{k-1}}{p_{k-1}},$$

i.e. $p_k = p_{k-1} + n_{k-1}, n_k = p_{k-1}$. Then we have on the one hand $p_1/n_1 = 2/1$, and on the other hand for all integers $k = 1, 2, 3, \ldots$

$$\frac{p_{k+1}}{n_{k+1}} = \frac{p_k + n_k}{p_k} = \frac{p_k + p_{k-1}}{p_{k-1} + n_{k-1}} = \frac{p_k + p_{k-1}}{n_k + n_{k-1}}.$$

This last formula detects the fraction p_{k+1}/n_{k+1} to be the mediant of the preceding fractions p_k/n_k and p_{k-1}/n_{k-1}. In other words: The sequence of fractions

$$\frac{p_0}{n_0} = \frac{1}{1}, \quad \frac{p_1}{n_1} = \frac{2}{1}, \quad \frac{p_2}{n_2} = \frac{3}{2}, \quad \frac{p_3}{n_3} = \frac{5}{3}, \quad \frac{p_4}{n_4} = \frac{8}{5}, \quad \ldots$$

form in the FAREY table the vertices of a zigzag-line, each vertex of this line being the mediant between the two former ones. The subsequence consisting of fractions p_k/n_k with even indices k is monotone increasing:

$$\frac{p_0}{n_0} = \frac{1}{1} < \frac{p_2}{n_2} = \frac{3}{2} < \frac{p_4}{n_4} = \frac{8}{5} < \frac{p_6}{n_6} = \frac{21}{13} < \frac{p_8}{n_8} = \frac{55}{34} < \ldots,$$

the subsequence consisting of fractions p_k/n_k with odd indices k is monotone decreasing:

$$\frac{p_1}{n_1} = \frac{2}{1} > \frac{p_3}{n_3} = \frac{5}{3} > \frac{p_5}{n_5} = \frac{13}{8} > \frac{p_7}{n_7} = \frac{34}{21} > \frac{p_9}{n_9} = \frac{89}{55} > \ldots,$$

and we always have $p_{2j+1}/n_{2j+1} > p_{2k}/n_{2k}$.

The relation $\varphi > 1 = p_0/n_0$ on the one hand, and the two formulae

$$\frac{p_{k+1}}{n_{k+1}} = 1 + \frac{1}{\frac{p_k}{n_k}} \quad \text{and} \quad \varphi = 1 + \frac{1}{\varphi}$$

on the other hand prove successively

$$\varphi < \frac{p_1}{n_1}, \quad \varphi > \frac{p_2}{n_2}, \quad \varphi < \frac{p_3}{n_3}, \quad \varphi > \frac{p_4}{n_4}, \quad \varphi < \frac{p_5}{n_5}, \quad \varphi > \frac{p_6}{n_6}, \quad \ldots.$$

Therefore φ always lies in between two consecutive fractions $p_k/n_k, p_{k+1}/n_{k+1}$ but never coincides with one of them. In intuitive words: the zigzag-line with p_k/n_k as vertices on the FAREY table meanders the value φ but it will never hit it. Thus φ is necessarily apart from any fraction of integers.

1.3 Continued fractions

Suppose that p/n and q/m denote consecutive fractions, say with p/n to the left of q/m, in a row of the FAREY table. We assert that the fractions

$$\frac{p+q}{n+m}, \quad \frac{2p+q}{2n+m}, \quad \frac{3p+q}{3n+m}, \quad \ldots, \quad \frac{kp+q}{kn+m}, \quad \ldots$$

are the first, the second, the third, ..., the k-th, ... mediant between p/n and q/m in the following rows that immediately adjoin to p/n on its right side. Indeed, this is true for $k = 1$, and if the assertion holds for k, then p/n and $(kp+q)/(kn+m)$ are consecutive fractions with p/n to the left of the fraction $(kp+q)/(kn+m)$ in the $(kn+m)$-th row of the FAREY table. The fact that the mediant of these fractions is

$$\frac{p+(kp+q)}{n+(kn+m)} = \frac{(k+1)p+q}{(k+1)n+m}$$

proves the assertion for $k+1$. In the same way the fractions

$$\frac{p+q}{n+m}, \quad \frac{p+2q}{n+2m}, \quad \frac{p+3q}{n+3m}, \quad \ldots, \quad \frac{p+kq}{n+km}, \quad \ldots$$

are the first, the second, the third, ..., the k-th, ... mediant between p/n and q/m in the following rows that immediately adjoin to q/m on its left side. With this perception in mind, a *continued fraction*

$$[a_0; a_1, a_2, \ldots, a_k, \ldots]$$

consisting of an integer a_0 and of a finite or an infinite sequence of positive integers a_1, a_2, ..., a_k, ... then is defined intuitively as a zigzag-line in the FAREY table that is constructed in the following way: It starts with the integer $p_0/n_0 = [a_0] = a_0$, i.e. with $p_0 = a_0$, $n_0 = 1$ as first and with the fraction

$$\frac{p_1}{n_1} = [a_0; a_1] = a_0 + \frac{1}{a_1} = \frac{a_1 a_0 + 1}{a_1}, \quad \text{i.e. with } p_1 = a_1 p_0 + 1, \, n_1 = a_1 n_0$$

as second vertex. If, for an arbitrary positive integer k, we already know the vertices, i.e. fractions

$$\frac{p_{k-1}}{n_{k-1}} = [a_0; a_1, \ldots, a_{k-1}], \quad \frac{p_k}{n_k} = [a_0; a_1, \ldots, a_{k-1}, a_k]$$

of the FAREY table, then the next vertex

$$\frac{p_{k+1}}{n_{k+1}} = [a_0; a_1, \ldots, a_{k-1}, a_k, a_{k+1}] = \frac{a_{k+1} p_k + p_{k-1}}{a_{k+1} n_k + n_{k-1}}$$

is the a_{k+1}-th mediant between p_{k-1}/n_{k-1} and p_k/n_k in the following rows that immediately adjoins to p_k/n_k. Otherwise stated: We can start with

$$p_{-2} = 0, \quad p_{-1} = 1, \quad n_{-2} = 1, \quad n_{-1} = 0,$$

and define inductively for $k = 0, 1, 2, 3, \ldots$

$$p_k = a_k p_{k-1} + p_{k-2}, \quad n_k = a_k n_{k-1} + n_{k-2}.$$

By doing this, we gain the fractions

$$\frac{p_0}{n_0} = [a_0], \quad \frac{p_1}{n_1} = [a_0; a_1], \quad \frac{p_2}{n_2} = [a_0; a_1, a_2],$$

$$\ldots, \quad \frac{p_k}{n_k} = [a_0; a_1, \ldots, a_{k-1}, a_k], \quad \ldots$$

in the FAREY table which form the vertices of the zigzag-line of the continued fraction $[a_0; a_1, a_2, \ldots, a_k, \ldots]$.

If the sequence of the positive integers $a_1, a_2, \ldots, a_k, \ldots$ terminates with the number a_j, then the last fraction

$$\frac{p_j}{n_j} = [a_0; a_1, \ldots, a_k, \ldots, a_j]$$

may be identified with the whole continued fraction itself. If the sequence of the positive integers $a_1, a_2, \ldots, a_k, \ldots$ is infinite, then the zigzag-line of the continued fraction dives into the FAREY table and will never touch down. The ratio of the diagonal to the lateral of the pentagram

$$\varphi = [1; 1, 1, 1, \ldots, 1, \ldots]$$

was our first example.

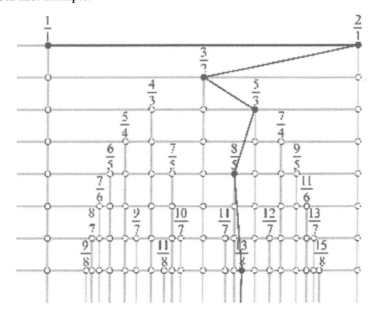

1.4 Special square roots

Let m denote a positive integer. There is a unique positive solution ψ of the equation

$$\psi = \frac{1}{2m + \psi}$$

which can be transformed to the quadratic equation $\psi^2 + 2m\psi - 1 = 0$ with the solution

$$\psi = \sqrt{m^2 + 1} - m \,.$$

The entity ψ allows to construct a continued fraction in total analogy to the ratio φ of the diagonal to the lateral of the pentagram:

Suppose we start with the fraction $p_0/n_0 = 0/1$, i.e. $p_0 = 0$, $n_0 = 1$, and we define by induction for $k = 1, 2, 3, \ldots$

$$\frac{p_k}{n_k} = \frac{1}{2m + \frac{p_{k-1}}{n_{k-1}}} = \frac{n_{k-1}}{2mn_{k-1} + p_{k-1}},$$

i.e. $p_k = n_{k-1}$, $n_k = 2mn_{k-1} + p_{k-1}$. Then we have on the one hand $p_1/n_1 = 1/2m$ and on the other hand for all integers $k = 1, 2, 3, \ldots$

$$\frac{p_{k+1}}{n_{k+1}} = \frac{n_k}{2mn_k + p_k} = \frac{2mn_{k-1} + p_{k+1}}{2mn_k + n_{k-1}} = \frac{2mp_k + p_{k-1}}{2mn_k + n_{k-1}} \,.$$

This last formula detects the fraction p_{k+1}/n_{k+1} to be the $2m$-th mediant between the preceding fractions p_k/n_k and p_{k-1}/n_{k-1} that immediately adjoins to p_k/n_k. In other words: The sequence of fractions

$$\frac{p_0}{n_0} = [0] \,, \quad \frac{p_1}{n_1} = [0; 2m] \,, \quad \frac{p_2}{n_2} = [0; 2m, 2m] \,,$$

$$\ldots, \quad \frac{p_k}{n_k} = [0; 2m, 2m, \ldots, 2m] \,, \quad \ldots$$

form in the FAREY table the vertices of a zigzag-line, each vertex of this line being the $2m$-th mediant between the two former ones. The subsequence consisting of fractions p_k/n_k with even indices k is monotone increasing:

$$\frac{p_0}{n_0} < \frac{p_2}{n_2} < \frac{p_4}{n_4} < \frac{p_6}{n_6} < \frac{p_8}{n_8} < \ldots,$$

the subsequence consisting of fractions p_k/n_k with odd indices k is monotone decreasing:

$$\frac{p_1}{n_1} > \frac{p_3}{n_3} > \frac{p_5}{n_5} > \frac{p_7}{n_7} > \frac{p_9}{n_9} > \ldots,$$

and we always have $p_{2j+1}/n_{2j+1} > p_{2k}/n_{2k}$.

The relation $\psi > 0 = p_0/n_0$ on the one hand and the two formulae

$$\frac{p_{k+1}}{n_{k+1}} = \frac{1}{2m + \frac{p_k}{n_k}} \quad \text{and} \quad \psi = \frac{1}{2m + \psi}$$

on the other hand prove successively

$$\psi < \frac{p_1}{n_1}, \quad \psi > \frac{p_2}{n_2}, \quad \psi < \frac{p_3}{n_3}, \quad \psi > \frac{p_4}{n_4}, \quad \psi < \frac{p_5}{n_5}, \quad \psi > \frac{p_6}{n_6}, \quad \ldots$$

Therefore ψ always lies in between two consecutive fractions p_k/n_k, p_{k+1}/n_{k+1} but never coincides with one of them. In intuitive words: the zigzag-line with p_k/n_k as vertices on the FAREY table meanders the value ψ but it will never hit it. Thus

$$\psi = [0; 2m, 2m, \ldots, 2m, \ldots] \quad \text{and} \quad \sqrt{m^2 + 1} = [m; 2m, 2m, \ldots, 2m, \ldots]$$

are necessarily apart from any fraction of integers. This seems to be the most intuitive proof for the fact that $\sqrt{2}, \sqrt{5}, \sqrt{10}, \sqrt{17}, \sqrt{26}, \ldots$ are irrational.

1.5 DEDEKIND cuts

All these foregoing results were at least implicitly known to the ancient Greek mathematicians. They, moreover, would hesitate to identify infinite continued fractions with geometrically or algebraically calculated "numbers" like φ or ψ: The ontological, or even better: the hermeneutic status of entities like φ or ψ in their eyes is completely vague. And indeed, it remained vague till the end of the nineteenth century when DEDEKIND once and for all believed to unveil the secret of these entities in his short booklet called *Continuity and Irrational Numbers* followed by a popular version called *What are and what should be Numbers* (written originally in German: *Was sind und was sollen die Zahlen*, published 1888 by Friedrich Vieweg & Sons).

DEDEKIND's visual idea was: to replace the zigzag-line of a continued fraction by a straight vertical line. If the continued fraction

$$[a_0; a_1, \ldots, a_k, \ldots, a_j]$$

terminates and defines a finite zigzag-line in the FAREY table which ends at

$$\frac{p_j}{n_j} = [a_0; a_1, \ldots, a_k, \ldots, a_j],$$

then DEDEKIND's straight line is that vertical line that hits the fraction p_j/n_j. If the continued fraction

$$[a_0; a_1, \ldots, a_k, \ldots]$$

does not terminate, then DEDEKIND's straight line is that vertical line that has all fractions of the form

$$\frac{p_{2j}}{n_{2j}} = [a_0; a_1, \ldots, a_k, \ldots, a_{2j}], \qquad (j = 0, 1, 2, \ldots),$$

on its left side and all fractions of the form

$$\frac{p_{2j+1}}{n_{2j+1}} = [a_0; a_1, \ldots, a_k, \ldots, a_{2j+1}], \qquad (j = 0, 1, 2, \ldots),$$

on its right side. In other words: The zigzag-line of the infinite continued fraction $[a_0; a_1, \ldots, a_k, \ldots]$ creeps along this straight line and intersects it after each vertex.

DEDEKIND transformed this visual idea into abstract mathematics with the help of CANTOR's set-theoretic language: A *cut* $\vartheta = (P|Q)$ is defined as a pair of two sets P, Q with the following properties:

1. Both, P and Q, are non-empty subsets of fractions of the FAREY table.
2. The union of P and Q is the whole FAREY table, i.e. the set of all rational numbers.
3. For all fractions $a = p/n$, $b = q/m$ of the FAREY table, the relations $a < b$ and $b \in P$ imply $a \in P$, and the relations $a < b$ and $a \in Q$ imply $b \in Q$.
4. The intersection of P and Q is either empty or consists of exactly one fraction.

Given the finite continued fraction

$$[a_0; a_1, \ldots, a_k, \ldots, a_j],$$

the DEDEKIND cut $(P|Q)$ defined by this continued fraction is easily described: The set P consists of all fractions

$$\frac{p}{n} \leq [a_0; a_1, \ldots, a_k, \ldots, a_j]$$

and the set Q consists of all fractions

$$\frac{q}{m} \geq [a_0; a_1, \ldots, a_k, \ldots, a_j].$$

Given the infinite continued fraction

$$[a_0; a_1, \ldots, a_k, \ldots],$$

the DEDEKIND cut $(P|Q)$ defined by this continued fraction is described as follows: Let r/l denote a fraction of the FAREY table. Suppose on the one hand that the positive integer k is big enough so that r/l can be found in the n_{2k}-th row (i.e. $n_{2k} \geq l$), then r/l is an element of P if and only if

$$\frac{r}{l} \leq \frac{p_{2k}}{n_{2k}} = [a_0; a_1, \ldots, a_{2k-1}, a_{2k}].$$

Suppose on the other hand that the positive integer k is big enough so that r/l can be found in the n_{2k+1}-th row (i.e. $n_{2k+1} \geq l$), then r/l is an element of Q if and only if
$$\frac{r}{l} \geq \frac{p_{2k+1}}{n_{2k+1}} = [a_0; a_1, \ldots, a_{2k}, a_{2k+1}].$$

It is exactly this definition which guarantees that the cut $(P|Q)$ is always between the left vertices
$$\frac{p_{2k}}{n_{2k}} = [a_0; a_1, \ldots, a_{2k-1}, a_{2k}], \qquad (k = 0, 1, 2, \ldots),$$

and the right vertices
$$\frac{p_{2k+1}}{n_{2k+1}} = [a_0; a_1, \ldots, a_{2k}, a_{2k+1}], \qquad (k = 0, 1, 2, \ldots),$$

of the zigzag-line which symbolizes $[a_0; a_1, \ldots, a_k, \ldots]$.

Insurmountable conceptual difficulties, however, are hidden behind the idea of DEDEKIND cuts. Take for instance the beginning of the continued fraction of π, the ratio between the circumference of the circle and its diameter:
$$\pi = [3; 7, 15, 1, \ldots].$$

As a matter of fact, we have
$$\pi > [3] = \frac{3}{1},$$
$$\pi < [3; 7] = \frac{7 \cdot 3 + 1}{7 \cdot 1 + 0} = \frac{22}{7},$$
$$\pi > [3; 7, 15] = \frac{15 \cdot 22 + 3}{15 \cdot 7 + 1} = \frac{333}{106},$$
$$\pi < [3; 7, 15, 1] = \frac{1 \cdot 333 + 22}{1 \cdot 106 + 7} = \frac{355}{113}.$$

But after that, we have to wait enormously long until we find the mediant between $333/106$ and $355/113$, which for the first time is less than π: Just the 292-nd mediant which immediately adjoins to $355/113$ on its left side does this job. It assures us
$$\pi > [3; 7, 15, 1, 292] \doteq \frac{292 \cdot 355 + 333}{292 \cdot 113 + 106} = \frac{103\,993}{33\,102},$$

but we have to dive to the 33102-nd row $\mathbb{Q}_{33102}$ of the FAREY table till we gain this result. And this example only indicates the subtlety that moves along with DEDEKIND cuts: There is no chance to be able to assure that each straight vertical line in the FAREY table divides the set of all fractions effectively into a cut $(P|Q)$. We even do not know this for the prominent example of the number π. And one has to abandon all hope that this could be done by factual calculation for the plethora of all possible straight vertical lines in the FAREY table.

Put in another way: DEDEKIND cuts represent sharp vertical lines of demarcation in the FAREY table whereas the initial segments

$$[a_0; a_1], \quad [a_0; a_1, a_2], \quad [a_0; a_1, a_2, a_3],$$

$$\ldots, \quad [a_0; a_1, \ldots, a_k], \quad \ldots$$

of an infinite continued fraction only define vertical stripes which, although they become thinner from step to step, never loose their width. Starting with the infinite continued fraction $[a_0; a_1, \ldots, a_k, \ldots]$, one has to accept the illusion of a synopsis of all infinite many stripes to gain the matching DEDEKIND cut by forming mentally their intersection. And starting with the DEDEKIND cut, one has to believe in the omniscience to decide for any given fraction of the FAREY table whether it lies on the right or on the left side of this cut in order to reconstruct the matching continued fraction $[a_0; a_1, \ldots, a_k, \ldots]$. HENRI POINCARE stressed that the set-theoretic foundation of these ideas should be criticized because of their lack of intuitive transparency – besides of the fact that naive set theory proved to be inconsistent, a fact of which BERTRAND RUSSELL and CANTOR himself became aware.

1.6 WEYL's alternative

HERMANN WEYL was convinced that a mathematical theory of the continuum not only has to be consistent, it must also be reasonable. He had come to believe that mathematical analysis at the beginning of the twentieth century would not bear logical scrutiny, for its essential concepts and procedures involved vicious circles to such an extent that "every cell (so to speak) of this mighty organism is permeated by contradiction." In his book *The Continuum*, written 1918, he proposed a new and constructive theory of the real numbers which satisfies consistency, and, as far as possible, intuitive transparency. But WEYL was doubtful as to the task that the idea of the continuum "given to us immediately by intuition (in the flow of time and of motion)" could be entirely transformed into a mathematical system: "The conceptual world of mathematics is so foreign to what the intuitive continuum presents to us that the demand for coincidence between the two must be dismissed as absurd."

In his book, however, WEYL tries to overcome possible vicious circles by providing analysis with a purely constructive formulation: He restricts analysis to what can be done in terms of integers with the aid of the basic logic operations, together with the operation of substitution and the process of "iteration", i.e. primitive recursion. In factual terms, this means that only those infinite continued fractions $[a_0; a_1, \ldots, a_k, \ldots]$ define admissible DEDEKIND cuts for which the assignments $0 \mapsto a_0, 1 \mapsto a_1, \ldots, k \mapsto a_k, \ldots$ are based on primitive recursive functions. Only these DEDEKIND cuts should constitute permissible real numbers.

WEYL recognized that the effect of this restriction on the one hand would be to render unprovable many of the central results of classical analysis – e.g. DIRICHLET's principle that any bounded non-empty set of real numbers has a least upper bound – but he was prepared to accept this as part of the price that must be paid for the security of mathematics. On the other hand, he saw that his construction did not provide an adequate representation of physical or temporal continuity as it is actually experienced. WEYL's substitute of the continuum is composed of individual real numbers which are well-defined and can be sharply distinguished. The result of his work is described by him in the following words: "An ensemble of individual points is, so to speak, picked out from the fluid paste of the continuum. The continuum is broken up into isolated elements, and the flowing-into-each-other of its parts is replaced by certain conceptual relations between these elements, based on the 'larger-smaller' relationship. This is why I speak of the *atomistic* conception of the continuum."

1.7 BROUWER's alternative

After the First World War, WEYL met LUITZEN EGBERTUS JAN BROUWER in Switzerland. DIRK VAN DALEN comments this meeting with the following words: "During the summer vacation BROUWER stayed in the Engadin, where WEYL visited him. With a little bit of luck the occasion could have been grander. HILBERT had visited Switzerland just before BROUWER, as it appears from a postcard from the latter to HILBERT, saying how sorry he was to have missed him. It would have been of decisive importance for the foundations of mathematics if the three of them could have sat down and discussed the new developments! HILBERT would have had the story of BROUWER's new intuitionism first hand, and it could have changed the course of the coming foundational conflict. These are history's little quirks."

The salient point of "BROUWER's new intuitionism" can be concentrated into one sentence: He replaced DEDEKIND cuts, i.e. vertical *lines* in the FAREY table, by vertical *stripes*.

In BROUWER's view, an "individual" real number is a sequence of overlapping stripes whose width converges to zero. An easy example is that of an infinite continued fraction $[a_0; a_1, \ldots, a_k, \ldots]$: the stripes here are bounded by the fractions

$$\frac{p_{2j}}{n_{2j}} = [a_0; a_1, \ldots, a_{2j-1}, a_{2j}] \quad \text{and} \quad \frac{p_{2j+1}}{n_{2j+1}} = [a_0; a_1, \ldots, a_{2j}, a_{2j+1}].$$

But the process that generates the integer a_0, and the positive integers a_1, a_2, ..., a_k, ... not in the least needs to be based on recursive functions, it can be totally arbitrary. What is more, BROUWER's sequences of stripes by no means are restricted to infinite continued fractions. It of course is possible, for any stripe of such a sequence, to construct a finite continued fraction $[a_0; a_1, \ldots, a_{2j}, a_{2j+1}]$

so that the stripe comprises both fractions

$$\frac{p_{2j}}{n_{2j}} = [a_0; a_1, \ldots, a_{2j-1}, a_{2j}] \quad \text{and} \quad \frac{p_{2j+1}}{n_{2j+1}} = [a_0; a_1, \ldots, a_{2j}, a_{2j+1}],$$

but this certainly does not guarantee the existence of an infinite continued fraction to which the sequence of stripes converges. So only the stripes themselves are at BROUWER's disposal, and he entirely repudiates the imagination, one could equally well calculate with their limit, i.e. with DEDEKIND cuts.

The personal contact between WEYL and BROUWER opened WEYL's eyes to, as he called it later, "revolutionary" views of BROUWER on the intuitive continuum. He welcomed the idea of sequences generated by free acts of choice, thus identifying the continuum as a "medium of free Becoming" which "does not dissolve into a set of real numbers as finished entities". WEYL felt that BROUWER, through his doctrine of intuitionism, had come closer than anyone else to bridging that "unbridgeable chasm" between the intuitive continuum and the mathematical continuum. He found the fact compelling that BROUWER's continuum cannot be split into two disjoint non-empty parts: "A genuine continuum," WEYL says, "cannot be divided into separate fragments." He even expresses this more colorfully by quoting ANAXAGORAS to the effect that a continuum "defies the chopping off of its parts with a hatchet."

In 1920 Weyl finished a provocative paper *On the New Foundational Crisis in Mathematics* (written originally in German: *Über die neue Grundlagenkrise der Mathematik*, published in *Mathematische Zeitschrift* **10**, 1921, 37-79). Here, after a revision of his own old ideas about the "predicative", atomistic continuum, WEYL gives an exposition of BROUWER's continuum with choice sequences and choice real numbers. In this paper, for the first time, the unsplittability of the continuum is announced, and, furthermore, WEYL asserts the most fundamental theorem of intuitionistic mathematics: "Above all, however, there can be no other functions at all on the continuum than continuous functions." Indeed, WEYL's paper reads as a manifesto to the mathematical community. It uses an evocative language as, for instance, the opening words demonstrate: "The antinomies of set theory are usually considered as border conflicts that concern only the remotest provinces of the mathematical empire and that can in no way imperil the inner solidity and security of the empire itself or of its genuine central areas. Almost all the explanations of these disturbances which were given by qualified sources (with the intention to deny them or to smooth them out), however, lack the character of a clear self-evident conviction, born of a totally transparent evidence, but belong to that way of half to three-quarters attempts at self-deception that one so frequently comes across in political and philosophical thought. Indeed, every earnest and honest reflection must lead to the insight that the troubles in the borderland of mathematics must be judged as symptoms, in which what lies hidden at the center of the superficially glittering and smooth activity comes to light – namely the inner instability of the foundations, upon which the structure of the empire rests."

But nearly at the same time, just few years later, WEYL suspected that the resulting gain in intuitive clarity had been bought at a considerable price, as witnessed by his remarks in the 1927 edition of *Philosophy of Mathematics and Natural Science*: "Mathematics with BROUWER gains its highest intuitive clarity. He succeeds in developing the beginnings of analysis in a natural manner, all the time preserving the contact with intuition much more closely than had been done before. It cannot be denied, however, that in advancing to higher and more general theories the inapplicability of the simple laws of classical logic eventually results in an almost unbearable awkwardness. And the mathematician watches with pain the greater part of his towering edifice which he believed to be built of concrete blocks dissolve into mist before his eyes."

In 1967, one year after BROUWER's death, ERRETT BISHOP demonstrated in his book *Foundations of Constructive Analysis* that a constructive treatment of analysis, even comprising intricate disciplines like functional analysis, is feasible. Although BISHOP, his colleague DOUGLAS S. BRIDGES, and the mathematicians of his American school of constructivism do not fully incorporate BROUWER's ideas, they at least could banish the fear of WEYL expressed in the quotation above.

One can, nevertheless, imagine that the majority of practising mathematicians have little patience with a doctrine that promises hard labor and fewer results. And this argument applies rightly BISHOP's approach to mathematics, but it utterly fails to BROUWER's intuitionism. Intuitionistic mathematics is not, as many believed, the loosely connected remnant of results of the powerful traditional mathematics. There is a good deal of structure, albeit more sophisticated than what traditional mathematicians were used to, as will be shown by the following example of higher analysis:

1.8 Integration in traditional and in intuitionistic framework

It is easy to explain the RIEMANN integrability within the unit interval $[0; 1]$ of bounded functions defined on sequences that are dense in $[0; 1]$. For instance the function f_λ with

$$f_\lambda\left(\frac{p}{n}\right) = \frac{1}{n^\lambda}$$

for all fractions p/n in the FAREY table is RIEMANN integrable for any positive λ with integral zero. The proof is easy: Given an arbitrary positive ε, the natural number m is constructed so large that $1/m^\lambda$ is less than $\varepsilon/2$. Now one fixes $\delta_\lambda = \varepsilon/2m^2$ to be an upper bound of the length $x_k - x_{k-1}$ of the intervals $[x_{k-1}; x_k]$ in the partition

$$0 = x_0 < x_1 < x_2 < \ldots < x_{j-1} < x_j = 1.$$

We now choose arbitrary rational intermediate points

$$0 \leq t_1 < t_2 < \ldots < t_j \leq 1$$

which obey for all $k = 1, 2, ..., j$ the condition $x_{k-1} \leq t_k \leq x_k$. As there are less than m^2 intervals $[x_{k-1}; x_k]$ containing a fraction p/n with denominator $n < m$, and because for all the other intervals $[x_{k-1}; x_k]$ the inequality $f_\lambda(t_k) < \varepsilon/2$ must hold, one gains

$$0 \leq \sum_{k=1}^{j} f_\lambda(t_k)(x_k - x_{k-1}) < m^2 \cdot 1 \cdot \delta_\lambda + \frac{\varepsilon}{2} \cdot 1 \leq \varepsilon$$

which confirms the integrability of f_λ with integral zero.

We therefore conclude on the one hand that

$$\lim_{\lambda \to 0+0} \int_0^1 f_\lambda = 0 \,.$$

This is not surprising for the traditional mathematician. In her or his eyes, f_λ is interpreted to be the function with the assignment $f_\lambda(x) = 1/n^\lambda$ if x coincides with the fraction p/n of the FAREY table, and with the assignment $f_\lambda(x) = 0$ if x is irrational. On the other hand we obtain

$$f_0 := \lim_{\lambda \to 0+0} f_\lambda$$

as function defined on the fractions p/n of the FAREY table with

$$f_0\left(\frac{p}{n}\right) = \lim_{\lambda \to 0+0} f_\lambda\left(\frac{p}{n}\right) = \lim_{\lambda \to 0+0} \frac{1}{n^\lambda} = 1 \,.$$

This function f_0, defined on the dense subset of rational numbers within the interval $[0; 1]$, is integrable too, but possesses 1 as value of the integral. This result deviates substantially from LEBESGUE's theorem about dominated convergence.

The reason of this deviation is clear from the viewpoint of the traditional mathematician: In her or his eyes, f_0 stands for the DIRICHLET function, attaining the value 1 on rational and the value 0 on irrational numbers. This function is of course not RIEMANN integrable but possesses a LEBESGUE integral with value zero, wholly in accordance with the dominated convergence theorem. One can even support this traditional argument by the following straightforward proof which uses a slight generalization of RIEMANN integrals, formulated by RALPH HENSTOCK about 50 years ago: A function f defined on $[0; 1]$ is *integrable*, fixing the real number

$$\int_0^1 f$$

as its integral, if there exists for any positive ε a *function δ* defined on $[0; 1]$ with *positive* values such that for any partition

$$0 = x_0 < x_1 < x_2 < ... < x_{j-1} < x_j = 1$$

of $[0; 1]$ and any choice of intermediate points

$$0 \leq t_1 < t_2 < ... < t_j \leq 1$$

which obey for all $k = 1, 2, ..., j$ the two conditions $x_{k-1} \leq t_k \leq x_k$ and $x_k - x_{k-1} \leq \delta(t_k)$, the inequality

$$\left| \sum_{k=1}^{j} f(t_k)(x_k - x_{k-1}) - \int_0^1 f \right| < \varepsilon$$

holds. Let for instance the sequence $(u_1, u_2, u_3, ..., u_n, ...)$ be an enumeration of the FAREY fractions between 0 and 1 and define the function δ by the formula $\delta(t) = \varepsilon/2^n$ in the case $t = u_n$, and $\delta(t) = 1$ otherwise. It is then very easy to check

$$\left| \sum_{k=1}^{j} f_\lambda(t_k)(x_k - x_{k-1}) \right| < \varepsilon \qquad \text{therefore} \qquad \int_0^1 f_\lambda = 0$$

for any $\lambda > 0$, and the fact that δ does not depend on the choice of λ especially proves

$$\left| \sum_{k=1}^{j} f_0(t_k)(x_k - x_{k-1}) \right| < \varepsilon \qquad \text{therefore} \qquad \int_0^1 f_0 = 0$$

with f_0 interpreted as DIRICHLET function – according to the dominated convergence theorem.

But this argument is by no means convincing in the eyes of BROUWER and WEYL: The reason why is that the above defined function δ is totally untenable within the intuitionistic theory. Any function defined on the whole interval [0; 1] must, according to an important theorem by BROUWER that will be discussed in subsection 4.2.4, be uniformly continuous there, and if the function is positive, it fixes a positive infimum. And indeed, in the constructive proof of the integrability of f_λ, the number δ_λ obviously depended on λ, and this forbids the limiting process $\lambda \to 0$ to be interchanged with the integral. Within intuitionism the generalization in the definition of RIEMANN integrals formulated by HENSTOCK does not extend the domain of RIEMANN integrable functions in the ordinary sense; differences between them and LEBESGUE, DENJOY, PERRON or HENSTOCK integrable functions disappear. Thus BROUWER's mathematics deviates substantially – and in a rather attractive way – from the broad pathway of traditional mathematics that relies on set-theory.

1.9 The wager

Nevertheless, GEORG POLYA, WEYL's colleague at Zürich, advanced the view that CANTOR's set theory should not be restricted, and heartily disagreed with WEYL's partisanship for BROUWER, as may be illustrated by the following small discussion:

POLYA: "You say that mathematical theorems should not only be correct, but also be meaningful. What is meaningful?"

1. Introduction and historical remarks

WEYL: "That is a matter of honesty."

POLYA: "It is error to mix philosophical statements in science. WEYL's continuum conception is emotion."

WEYL: "What POLYA calls emotion and rhetoric, I call insight and truth; what he calls science, I call symbol-pushing. POLYA's defence of set theory (that one may one day provide meaning to these formulations) is mysticism. To separate mathematics, as being formal, from spiritual life, kills it, turns it into a shell. To say that only the chess game is science, and that insight is not, *that* is a restriction."

The culmination point of the debate between POLYA and WEYL was the famous wager that was arranged on February 1918:

"Within 20 years POLYA and the majority of representative mathematicians will admit that the statements
1) *Every bounded set of reals has a precise supremum*
2) *Every infinite set of numbers contains a denumerable subset*
contain totally vague concepts, such as 'number', 'set' and 'denumerable', and that therefore their truth or falsity has the same status as that of the main propositions of HEGEL's natural philosophy.

However, under a natural interpretation 1) and 2) will be seen to be false."

When the bet was called, everyone agreed that POLYA had won with the remarkable exception of KURT GOEDEL. Although even WEYL did admit losing the bet, the "riddle of the continuum" retained its fascination for him: In one of his last papers, written in 1954, we find the observation that "the constructive transition to the continuum of real numbers is a serious matter (...) and I am bold enough to say that not even to this day are the logical issues involved in that constructive concept completely clarified and settled." It remains noteworthy that the construction of the real numbers contains subtleties that troubled such an acute intellect as GOEDEL's in 1940. But, oddly enough, the debate on the foundational crisis died away, probably also on account of the political cataclysm that during this time threatened the world and damaged the scientific climate in Europe.

The extraordinary history concerning the construction of real numbers, with CANTOR, DEDEKIND, BROUWER, WEYL and many others as its heroes, is most probably unparalleled in the whole universe. It, at least, would be tempting to renew the wager between POLYA and WEYL in the following form: If, in future times, we should succeed to get in contact with intelligent creatures in outer space, say on the asteroid B612, and if these creatures have developed mathematics to that extent that they have equivalent formulations of theorems like the *theorema aureum* in number theory or the *theorema egregium* in differential geometry, then we should not be surprised that this civilization, confronted with the wager between POLYA and WEYL, would be in favour of WEYL and would understand the theoretical development of the continuum according to the following lines.

1.10 How to read the following pages

The following text, condensed into one metaphor, presents a *smooth, straight highway* to the real numbers.

In order to gain the continuum, analysis courses usually favour either the axiomatic or the constructive method. We exclude the first from the following discussion: it does not offer any "way" to the continuum, it represents – again metaphorically speaking – rather an arrival without departure. Moreover, the axiomatic setting tends to hide the intrinsic nature of the real number system.

Two ways to construct the continuum nowadays are in common use: the first is paved with DEDEKIND cuts and the second with CAUCHY sequences. However, we suggest a third approach, which is founded on decimal expansions. This preference is proposed for three reasons:

First, it is a very *natural* way (hence "smooth"): Decimal expansion is the most common method of fixing a number – a paradigmatic example is the famous quest of BORWEIN et al. to see what the sequence of the first million, billion, trillion, ... decimal places of π might look like – and it is based only on the arithmetic of the integers *without any prerequisite of rational numbers*. As will be shown in section 3.1, an approach to the continuum using CAUCHY sequences is equally well possible (so one could for instance define real numbers via infinite sequences of finite continued fractions).

Second, it is a very *constructive* way (hence "straight"): Although this text is indebted to BROUWER's intuitionistic mathematics, most of it can equally well be read from the point of view of Russian recursive, as well as from that of BISHOP's constructive mathematics, and of course from that of formal mathematics in the tradition of CANTOR and DEDEKIND. Only the theorems in subsections 3.2.8, 4.1.4, 4.2.4, and 4.4.2 are typical of intuitionism and in variance with traditional mathematics or with recursive constructivism.

Third, it is a comparatively *short* way (hence "highway"): To reach the setting of the real number system is a straightforward procedure of few lemmas and theorems with non-trivial proofs on the one hand, and of several propositions that are almost obvious and corollaries that follow immediately on the other. However, and this should be considered as an additional advantage: each tiny step is proved in this text and nothing is – in contrast to many introductions to the real number system – "left as exercise to the reader". This – at first glance – tedious rigor turns out to be necessary in order to highlight the depth of the concepts being introduced.

2
Real numbers

2.1 Definition of real numbers

2.1.1 Decimal numbers

"The whole numbers are created by God, and everything else is the work of man." KRONECKER is right indeed: nothing is more fundamental than the concept of the numbers $1, 2, 3, \ldots, n, n+1, \ldots$. Very simple constructions, for instance formal differences $n - m$, uniquely determined by setting

$$n - m = \begin{cases} k & \text{if } n > m \text{ and } n - m = k \\ 0 & \text{if } n = m \\ -l & \text{if } n < m \text{ and } m - n = l \end{cases}$$

lead to the integers $0, 1, -1, 2, -2, \ldots, n, -n, \ldots$. These objects together with the elementary calculations of addition, subtraction, multiplication can be supposed as well known. The order relation, expressed by

$$\ldots < -n-1 < -n < \ldots < -2 < -1 < 0 < 1 < 2 < \ldots < n < n+1 < \ldots$$

immediately leads to a commonly used sensual concept: Integers are visualized as points on a straight line such that, for all integers p, the points $p-1$, p and the points p, $p+1$ have the same positive distance. This picture immediately raises the question how to handle mathematically the points of the line *between* two adjacent integers. A useful proposal is to fill these gaps by zooming the line by a positive factor greater than one – we in the following decide to take the factor 10 – and to copy the points of the original line on the stretched one. In other words: the unit 1 is replaced by the unit 1/10. An iteration of this zooming-process finally leads to the concept of decimal numbers:

We define a *decimal number* a as a pair $a = p \times 10^r$ of two integers p and r. The *equality*
$$p \times 10^r = q \times 10^s$$
holds in the case $s \geq r$ if and only if $p = q \times 10^{s-r}$, and holds in the case $s \leq r$ if and only if $p \times 10^{r-s} = q$. Given a decimal number $a = q \times 10^s$ and an arbitrary integer $r \leq s$, the number can also be represented as $a = p \times 10^r$ by setting $p = q \times 10^{s-r}$. *Addition, multiplication,* and *order* are defined by setting
$$p \times 10^r + q \times 10^r = (p+q) \times 10^r,$$
$$(p \times 10^r)(q \times 10^s) = (pq) \times 10^{r+s},$$
and $p \times 10^r > 0$ if and only if p is positive. *Division by* 2 *or* 5 is defined as
$$(p \times 10^r)/2 = 5p \times 10^{r-1}, \quad (p \times 10^r)/5 = 2p \times 10^{r-1}.$$
This, of course, generalizes to
$$(p \times 10^r)/2^m 5^n = (2^{\max(m,n)-m} 5^{\max(m,n)-n} p) \times 10^{r-\max(m,n)}.$$
These very elementary relations and operations transfer the arithmetic of integers to that of decimal numbers.

It is evident that decimal numbers do not constitute the full line: each zooming process, iterated arbitrarily often, will always leave gaps between two adjacent decimal numbers. The line itself appears as an unreachable background of all these distinct points. To analyze this background mathematically and to convert the concrete picture of the line into the abstract concept of the *continuum* is the content of this book.

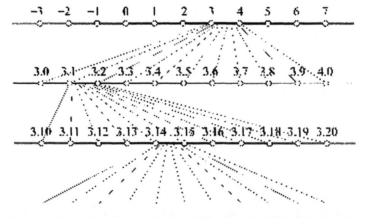

The letters a, b, c, d, e are used in the following for (finite) decimal numbers, the letters j, k, l, m, n are used for positive integers and the letters p, q, r, s are used for integers (positive, zero or negative).

2.1.2 Rounding of decimal numbers

For any integer $n > 0$ a decimal number *a with exactly n decimal places* is a number of the form

$$a = z + 0.z_1 z_2 \ldots z_n = z + z_1 \times 10^{-1} + z_2 \times 10^{-2} + \ldots + z_n \times 10^{-n}$$

z being an integer and $z_1, z_2, \ldots, z_n$ being digits, i.e. integers between 0 and 9.

Of course, a can be written in the form $a = p \times 10^{-n}$ with

$$p = 10^n z + 10^{n-1} z_1 + 10^{n-2} z_2 + \ldots + z_n .$$

The possibility $z_n = 0$ is *not* excluded.

Rounding of decimal numbers. *For each decimal number a with exactly $n+k$ decimal places there exists a uniquely determined decimal number $\{a\}_n$ with exactly n decimal places such that*

$$|a - \{a\}_n| \leq 5 \times 10^{-n-1} ,$$

and in the case $|a - \{a\}_n| = 5 \times 10^{-n-1}$ *one has*

$$\{a\}_n = \begin{cases} a + 5 \times 10^{-n-1} & \text{if } a > 0, \\ a - 5 \times 10^{-n-1} & \text{if } a < 0. \end{cases}$$

Proof. If $a = z + 0.z_1 z_2 \ldots z_n z_{n+1} \ldots z_{n+k}$ we fix

$$\{a\}_n = z + z_1 \times 10^{-1} + z_2 \times 10^{-2} + \ldots + z_n \times 10^{-n}$$

resp.

$$\{a\}_n = z + z_1 \times 10^{-1} + z_2 \times 10^{-2} + \ldots + z_n \times 10^{-n} + 10^{-n},$$

according to

$$z_{n+1} \times 10^{-n-1} + \ldots + z_{n+k} \times 10^{-n-k} < 5 \times 10^{-n-1}$$

resp.

$$z_{n+1} \times 10^{-n-1} + \ldots + z_{n+k} \times 10^{-n-k} > 5 \times 10^{-n-1} ,$$

and in the remaining case

$$z_{n+1} \times 10^{-n-1} + \ldots + z_{n+k} \times 10^{-n-k} = 5 \times 10^{-n-1} ,$$

i.e. $z_{n+1} = 5$ and $z_{n+2} = \ldots = z_{n+k} = 0$, according to

$$a < 0 \quad \text{resp.} \quad a > 0 .$$

It is obvious that this construction yields the desired result and that $\{a\}_n$ is uniquely determined by the conditions mentioned above. ∎

For any decimal number a with exactly $n + k$ decimal places we have

$$\{-a\}_n = -\{a\}_n .$$

Further the relations $a \geq 0$ and $\{a\}_n \geq 0$ are equivalent, as well as the relations $a \leq 0$ and $\{a\}_n \leq 0$.

2.1.3 Definition and examples of real numbers

A *real number* α is a sequence of decimal numbers

$$[\alpha]_1, [\alpha]_2, \ldots, [\alpha]_n, \ldots$$

with the conditions that for each n the element $[\alpha]_n$ is a decimal number with exactly n decimal places and that for each n and each m the inequalities

$$|[\alpha]_n - [\alpha]_m| \leq 10^{-n} + 10^{-m}$$

hold.

The letters $\alpha, \beta, \gamma, \delta, \varepsilon$ are used in the following for real numbers.

An *infinite decimal number* α is a sequence of decimal numbers

$$[\alpha]_1, [\alpha]_2, \ldots, [\alpha]_n, \ldots$$

with the conditions that for each n the element $[\alpha]_n$ is a decimal number with exactly n decimal places and that for each n and each k the decimal number $[\alpha]_{n+k}$ is of the form

$$[\alpha]_n + 0.00\ldots 0 z_{n+1} \ldots z_{n+k}$$

(with at least n zeros after the decimal point).

Each infinite decimal number is a real number.

Proof. Without loss of generality $m = n + k$, therefore

$$\left|[\alpha]_n - [\alpha]_{n+k}\right| = 0.00\ldots 0 z_{n+1} \ldots z_{n+k} \leq 10^{-n} \leq 10^{-n} + 10^{-n-k}$$

for any n and k. ∎

Each decimal number can uniquely be identified with an infinite decimal number.

Proof. Let $a = z + 0.z_1 z_2 \ldots z_m$ be a decimal number with exactly m decimal places. The sequence of numbers $[a]_1, [a]_2, \ldots, [a]_n, \ldots$ with

$$[a]_n = \begin{cases} z + 0.z_1 z_2 \ldots z_n & \text{if } n \leq m \\ z + 0.z_1 z_2 \ldots z_m 0 \ldots 0 & \text{if } n > m \end{cases}$$

(with $n-m$ zeros after z_m in the second case) clearly is an infinite decimal number.
∎

"Par abus de langage" we indicate the infinite decimal number produced by this proof by the same letter a as the decimal number itself.

As a first nontrivial example of a real number, let a denote a positive decimal number. There exists one and only one integer $z \geq 0$ such that

$$z^2 \leq a \quad \text{but} \quad (z+1)^2 > a.$$

Suppose for the positive integer n that we already have calculated digits z_1, z_2, ..., z_{n-1}. Then there also exists a uniquely defined digit z_n such that

$$(z + 0.z_1 z_2 \ldots z_{n-1} z_n)^2 \leq a \quad \text{but} \quad (z + 0.z_1 z_2 \ldots z_{n-1} z_n + 10^{-n})^2 > a \,.$$

The infinite decimal number $\sqrt{a}$ finally is defined by the formula

$$[\sqrt{a}]_n = z + 0.z_1 z_2 \ldots z_n$$

for all positive integers n. Of course, we are still far removed from a proof that $\sqrt{a}^2 = a$. We do not yet know how to multiply two real numbers, or even how equality of real numbers is defined. All this will be elaborated step by step on the following pages.

A *pendulum number* α is a sequence of decimal numbers

$$[\alpha]_1, [\alpha]_2, \ldots, [\alpha]_n, \ldots$$

with the conditions that for each n the element $[\alpha]_n$ is a decimal number with exactly n decimal places and that for each n and each k the decimal number $[\alpha]_{n+k}$ is of the form

$$[\alpha]_{n+k} = [\alpha]_n + w_{n+1} \times 10^{-n-1} + \ldots + w_{n+k} \times 10^{-n-k}$$

the $w_{n+1}, \ldots, w_{n+k}$ being integers with $-9 \leq w_{n+j} \leq 9$ for all $j = 1, \ldots, k$.

Each pendulum number is a real number.

Proof. Without loss of generality $m = n + k$, therefore

$$\begin{aligned} |[\alpha]_n - [\alpha]_{n+k}| &\leq |w_{n+1}| \times 10^{-n-1} + \ldots + |w_{n+k}| \times 10^{-n-k} \\ &\leq 9 \times 10^{-n-1} + \ldots + 9 \times 10^{-n-k} \\ &\leq 10^{-n} \leq 10^{-n} + 10^{-n-k} \end{aligned}$$

for any n and k. ∎

A second nontrivial example of a real number is a "black-box-number". This is a pendulum number α that is constructed in the following way: We type an arbitrary positive integer n as input into a machinery, called "black box", which with this input generates an integer w_n as output that is uniquely assigned by the input n and fulfills $-9 \leq w_n \leq 9$. In order to obtain for a positive integer n the number $[\alpha]_n$, one has to type the inputs $1, 2, \ldots, n$ into the black box, to collect the according outputs $w_1, w_2, \ldots, w_n$ and to define

$$[\alpha]_n = w_1 \times 10^{-1} + w_2 \times 10^{-2} + \ldots + w_n \times 10^{-n}.$$

We do not exclude black-box-numbers for which the black box may obey rather peculiar devices, for instance:

1. it announces the output $w_n = 5$ if the digits $z_n, z_{n+1}, \ldots, z_{2n}$ of $\left[\sqrt{2}\right]_{2n}$ coincide, i.e. $z_n = z_{n+1} = \ldots = z_{2n}$, and are equal to an even number, it announces the output $w_n = -5$ if the digits $z_n, z_{n+1}, \ldots, z_{2n}$ of $\left[\sqrt{2}\right]_{2n}$ coincide, i.e. $z_n = z_{n+1} = \ldots = z_{2n}$, and are equal to an odd number, and it announces the output $w_n = 0$ in any other case,

or

2. it throws a die, and announces the output $w_n = 5$ if the outcomes $z_n, z_{n+1}, \ldots, z_{2n}$ of the n-th, $(n+1)$-th, $\ldots$, $(2n)$-th cast coincide, i.e. $z_n = z_{n+1} = \ldots = z_{2n}$, and are equal to an even number, it announces the output $w_n = -5$ if the outcomes $z_n, z_{n+1}, \ldots, z_{2n}$ of the n-th, $(n+1)$-th, $\ldots$, $(2n)$-th cast coincide, i.e. $z_n = z_{n+1} = \ldots = z_{2n}$, and are equal to an odd number, and it announces the output $w_n = 0$ in any other case.

2.1.4 Differences and absolute differences

The *difference* $\alpha - \beta$ of two real numbers α and β is the sequence of decimal numbers $[\alpha - \beta]_1, [\alpha - \beta]_2, \ldots, [\alpha - \beta]_n, \ldots$ that are defined as

$$[\alpha - \beta]_n = \{[\alpha]_{n+1} - [\beta]_{n+1}\}_n .$$

The difference of two real numbers is a real number.

Proof. As

$$|[\alpha - \beta]_n - ([\alpha]_{n+1} - [\beta]_{n+1})| \leq 5 \times 10^{-n-1}$$

holds for any integer $n > 0$, the inequality

$$\begin{aligned}
|[\alpha - \beta]_n - [\alpha - \beta]_m| &\leq |[\alpha - \beta]_n - ([\alpha]_{n+1} - [\beta]_{n+1})| \\
&\quad + |[\alpha]_{n+1} - [\alpha]_{m+1}| + |[\beta]_{m+1} - [\beta]_{n+1}| \\
&\quad + |([\alpha]_{m+1} - [\beta]_{m+1}) - [\alpha - \beta]_m| \\
&\leq 5 \times 10^{-n-1} + 10^{-n-1} + 10^{-m-1} \\
&\quad + 10^{-n-1} + 10^{-m-1} + 5 \times 10^{-m-1} \\
&\leq 10^{-n} + 10^{-m}
\end{aligned}$$

follows for any n and m. ■

The *absolute difference* $|\alpha - \beta|$ of two real numbers α and β is the sequence of decimal numbers $[|\alpha - \beta|]_1, [|\alpha - \beta|]_2, \ldots, [|\alpha - \beta|]_n, \ldots$ that are defined as

$$[|\alpha - \beta|]_n = |[\alpha - \beta]_n|$$

For all real numbers α, β

$$[|\alpha - \beta|]_n = \begin{cases} [\alpha - \beta]_n & \text{if } [\alpha - \beta]_n \geq 0 \\ [\beta - \alpha]_n & \text{if } [\alpha - \beta]_n \leq 0 \end{cases}$$

Proof. The chain of equalities

$$\begin{aligned}[\beta - \alpha]_n &= \{[\beta]_{n+1} - [\alpha]_{n+1}\}_n = \{-([\alpha]_{n+1} - [\beta]_{n+1})\}_n \\ &= -\{[\alpha]_{n+1} - [\beta]_{n+1}\}_n = -[\alpha - \beta]_n\end{aligned}$$

proves this formula. ∎

The absolute difference of two real numbers is a real number.

Proof. This follows from the fact that $||a| - |b|| \leq |a - b|$ holds for any pair of decimal numbers. ∎

2.2 Order relations

2.2.1 Definitions and criteria

The *strict order* $\alpha > \beta$ signifies for real numbers α, β that one can find an integer $n > 0$ with the property $[\alpha - \beta]_n > 10^{-n}$. The notations $a > \beta$ and $\beta < \alpha$ are equivalent.

The *weak order* $\alpha \leq \beta$ signifies for real numbers α, β that one can prove $[\alpha - \beta]_n \leq 10^{-n}$ for all integers $n > 0$. The notations $a \leq \beta$ and $\beta \geq \alpha$ are equivalent.

The indirect proof. *A contradiction of the assumption $\alpha > \beta$ proves the inequality $\alpha \leq \beta$.*

Proof. For any given integer $n > 0$ the assumption $[\alpha - \beta]_n > 10^{-n}$ must be false, and this proves the inequality $[\alpha - \beta]_n \leq 10^{-n}$. ∎

However, we strictly *avoid* deriving $\alpha > \beta$ from the fact that $\alpha \leq \beta$ leads to a contradiction for the following reason: The impossibility of $[\alpha - \beta]_n \leq 10^{-n}$ for all integers $n > 0$ in general *does not* show how to pick up an integer $n > 0$ that guarantees $[\alpha - \beta]_n > 10^{-n}$.

Criterion of strict order. *The strict order $\alpha > \beta$ holds for real numbers α, β if and only if one can find a positive decimal number d and a positive integer j such that for all integers $n \geq j$ the inequality $[\alpha]_n > [\beta]_n + d$ follows.*

Proof. On the one hand, suppose that $\alpha > \beta$. By definition there exists an integer $m > 0$ with the property $[\alpha - \beta]_m > 10^{-m}$ and therefore

$$[\alpha]_{m+1} - [\beta]_{m+1} > 5 \times 10^{-m-1} .$$

Define
$$d = \frac{[\alpha]_{m+1} - [\beta]_{m+1}}{5} - 10^{-m-1},$$

and define the positive integer j so big that $10^{-j} \leq 2d$ is guaranteed. Then for all integers $n \geq j$ the desired inequality follows from

$$\begin{aligned}
[\alpha]_n - [\beta]_n &\geq [\alpha]_{m+1} - [\beta]_{m+1} \\
&\quad - |([\alpha]_{m+1} - [\beta]_{m+1}) - ([\alpha]_n - [\beta]_n)| \\
&\geq [\alpha]_{m+1} - [\beta]_{m+1} \\
&\quad - |[\alpha]_{m+1} - [\alpha]_n| - |[\beta]_n - [\beta]_{m+1}| \\
&\geq [\alpha]_{m+1} - [\beta]_{m+1} - 2 \times \left(10^{-m-1} + 10^{-n}\right) \\
&> \left([\alpha]_{m+1} - [\beta]_{m+1} - 5 \times 10^{-m-1}\right) - 2 \times 10^{-j} \\
&\geq 5d - 4d = d.
\end{aligned}$$

On the other hand, suppose that one can find a positive decimal number d and an integer $j > 0$ such that for all integers $n \geq j$ the inequality $[\alpha]_n > [\beta]_n + d$ holds: Choose the integer $n \geq j$ so big that $3 \times 10^{-n} \leq 2d$ is guaranteed. From the inequality

$$[\alpha]_{n+1} - [\beta]_{n+1} > d \geq 15 \times 10^{-n-1} = 10^{-n} + 5 \times 10^{-n-1}$$

we deduce that $[\alpha - \beta]_n > 10^{-n}$, and from that the relation $\alpha > \beta$ follows by definition. ∎

Criterion of weak order. *The weak order $\alpha \leq \beta$ holds for real numbers α, β if and only if for all positive decimal numbers e one can find a positive integer j such that for all integers $n \geq j$ the inequality $[\alpha]_n < [\beta]_n + e$ follows.*

Proof. On the one hand, suppose that $\alpha \leq \beta$: Choose for any positive decimal number $e > 0$ the integer $j > 1$ to so big that $15 \times 10^{-j} < e$ is guaranteed. The relation

$$[\alpha - \beta]_{n-1} \leq 10^{-(n-1)}$$

implies that for all integers $n \geq j$

$$[\alpha]_n - [\beta]_n \leq [\alpha - \beta]_{n-1} + 5 \times 10^{-n} \leq 15 \times 10^{-n} \leq 15 \times 10^{-j}$$

and thus the desired inequality $[\alpha]_n < [\beta]_n + e$.

On the other hand, suppose that for all positive decimal numbers d one can find a positive integer j such that for all integers $n \geq j$ the inequality $[\alpha]_n < [\beta]_n + d$ is correct: By the above lemma the possibility $\alpha > \beta$ is ruled out and we therefore have $\alpha \leq \beta$. ∎

2.2.2 Properties of the order relations

For all real numbers α, β, γ:
1. $\alpha > \beta$ together with $\beta > \alpha$ is absurd.
2. $\alpha > \alpha$ is absurd.
3. $\alpha > \beta$ implies $\alpha \geq \beta$.
4. $\alpha > \beta$ together with $\beta \geq \gamma$ implies $\alpha > \gamma$.
5. $\alpha \geq \beta$ together with $\beta > \gamma$ implies $\alpha > \gamma$.
6. $\alpha \leq \beta$ together with $\beta \leq \gamma$ implies $\alpha \leq \gamma$.

Proof. 1: $\alpha > \beta$ implies the existence of a positive d_1 and an integer j_1 such that

$$[\alpha]_n > [\beta]_n + d_1$$

for all $n \geq j_1$. $\beta > \alpha$ implies the existence of a positive d_2 and an integer j_2 such that

$$[\beta]_n > [\alpha]_n + d_2$$

for all $n \geq j_2$. This, of course, is absurd for $n \geq \max(j_1, j_2)$.

2: $\alpha > \alpha$ implies the existence of a positive decimal number d and a positive integer j such that for all integers $n \geq j$ we would have $[\alpha]_n > [\alpha]_n + d$ which (for $n = j$) is impossible.

3. As $\alpha > \beta$ forbids $\beta > \alpha$, we conclude $\beta \leq \alpha$.

4: $\alpha > \beta$ implies the existence of a positive d' and an integer j_1 such that

$$[\alpha]_n > [\beta]_n + d'$$

for all $n \geq j_1$. $\gamma \leq \beta$ implies the existence of an integer j_2 such that

$$[\gamma]_n < [\beta]_n + \frac{d'}{2}$$

for all $n \geq j_2$. Defining $d = d'/2$, $j = \max(j_1, j_2)$ we thus conclude

$$[\alpha]_n > [\gamma]_n + d$$

for all $n \geq j$.

5: $\beta > \gamma$ implies the existence of a positive d' and an integer j_1 such that

$$[\beta]_n > [\gamma]_n + d'$$

for all $n \geq j_1$. $\beta \leq \alpha$ implies the existence of an integer j_2 such that

$$[\beta]_n < [\alpha]_n + \frac{d'}{2}$$

for all $n \geq j_2$. Defining $d = d'/2$, $j = \max(j_1, j_2)$ we thus conclude

$$[\alpha]_n > [\gamma]_n + d$$

for all $n \geq j$.

6: Let e be an arbitrary positive decimal number. $\alpha \leq \beta$ implies the existence of an integer j_1 such that

$$[\alpha]_n < [\beta]_n + \frac{e}{2}$$

for all $n \geq j_1$. $\beta \leq \gamma$ implies the existence of an integer j_2 such that

$$[\beta]_n < [\gamma]_n + \frac{e}{2}$$

for all $n \geq j_2$. We thus conclude

$$[\alpha]_n < [\gamma]_n + e$$

for all $n \geq \max(j_1, j_2)$. ∎

From now on we may write down *chains* of inequalities like

$$\alpha < \beta < \gamma \text{ or } \alpha \leq \beta < \gamma \text{ or } \alpha < \beta \leq \gamma \text{ or } \alpha \leq \beta \leq \gamma.$$

The weak order $\alpha \leq \beta$ holds if and only if for all real numbers γ the strong order $\gamma > \beta$ implies the strong order $\gamma > \alpha$.

Proof. First assume that $\gamma > \beta$ implies $\gamma > \alpha$. The hypothesis $\alpha > \beta$ therefore would lead to the contradiction $\alpha > \alpha$. Thus $\alpha \leq \beta$ must hold.
The converse is the same statement as point 4 in the above corollary. ∎

For all decimal numbers a, b:
1. $a > b$ defined within the system of real numbers is equivalent to $a > b$ defined within the system of decimal numbers.
2. $a \leq b$ defined within the system of real numbers is equivalent to $a \leq b$ defined within the system of decimal numbers.

Proof. Assume $a = p \times 10^{-m}$, $b = q \times 10^{-m}$. Without loss of generality, m can be assumed to be a positive integer.

1: $a > b$ defined within the system of decimal numbers implies that $p \geq q + 1$. We set $j = m$, $d = 10^{-m-1}$ and obtain for all $n \geq j$

$$[a]_n = a = p \times 10^{-m} > q \times 10^{-m} + d = b + d = [b]_n + d$$

and therefore $a > b$ defined within the system of real numbers. On the other hand $a > b$ defined within the system of real numbers implies for a sufficiently big $n \geq m$ at least that $[a]_n > [b]_n$ which immediately leads to $a > b$ defined within the system of decimal numbers.

2: As $a \leq b$ holds within the system of decimal numbers if and only if for all decimal numbers c the strong order $c > b$ implies the strong order $c > a$, this equivalence is a direct consequence from point 1. ∎

For all decimal numbers a, b (with $a+(-b)$ resp. $a-b$ designing the difference within the system of decimal resp. of real numbers):

1. $a - b \leq a + (-b) \leq a - b$.
2. $|a - b| \leq |a + (-b)| \leq |a - b|$.

Proof. Assume $a = p \times 10^{-m}$, $b = q \times 10^{-m}$. Without loss of generality, m can be assumed to be a positive integer. Then $n \geq m$ implies that $[a]_n = a$, $[b]_n = b$. This proves $[a]_{n+1} - [b]_{n+1}$ being a decimal number with at most n decimal places, i.e. rounding this number to n decimal places does not change it. Therefore $n \geq m$ implies that

$$[a - b]_n = [a + (-b)]_n \;,$$

with the consequence

$$[a - b]_n \leq [a + (-b)]_n \leq [a - b]_n \;,$$

and

$$[|a - b|]_n = |[a - b]_n| \leq \big|[a + (-b)]_n\big| \leq |[a - b]_n| = [|a - b|]_n \;.$$

∎

2.2.3 Order relations and differences

The relations $\alpha > \beta$, $\alpha - \gamma > \beta - \gamma$, and $\gamma - \beta > \gamma - \alpha$ are equivalent.

Proof. The inequality $\alpha > \beta$ implies the existence of a positive d_1 and an integer j_1 such that $[\alpha]_n > [\beta]_n + d_1$ for all $n \geq j_1$. Define $d = d_1/2$ and the integer $j \geq j_1$ so big that $10^{-j} \leq d$. For all $n \geq j$ we thus conclude

$$\begin{aligned}
[\alpha - \gamma]_n &= \{[\alpha]_{n+1} - [\gamma]_{n+1}\}_n \geq [\alpha]_{n+1} - [\gamma]_{n+1} - 5 \times 10^{-n-1} \\
&> [\beta]_{n+1} - [\gamma]_{n+1} - 5 \times 10^{-n-1} + d_1 \\
&\geq \{[\beta]_{n+1} - [\gamma]_{n+1}\}_n - 10^{-n} + d_1 \\
&\geq [\beta - \gamma]_n - 10^{-j} + 2d \geq [\beta - \gamma]_n + d \;.
\end{aligned}$$

This proves $\alpha - \gamma > \beta - \gamma$.

The inequality $\alpha - \gamma > \beta - \gamma$ implies the existence of a positive d and an integer j such that $[\alpha - \gamma]_n > [\beta - \gamma]_n + d$ for all $n \geq j$. This implies for all $n > j$ that

$$[\beta - \gamma]_n + d = -[\gamma - \beta]_n + d < [\alpha - \gamma]_n = -[\gamma - \alpha]_n$$

with the conclusion $[\gamma - \beta]_n > [\gamma - \alpha]_n + d$. This proves $\gamma - \beta > \gamma - \alpha$.

The inequality $\gamma - \beta > \gamma - \alpha$ implies the existence of a positive d_1 and an integer j_1 such that $[\gamma - \beta]_n > [\gamma - \alpha]_n + d_1$ for all $n \geq j_1$. Define $d = d_1/2$ and the integer $j > j_1$ so big that $10^{-j+1} \leq d$. For all $n \geq j$ we thus conclude that

$$\begin{aligned}
\{[\gamma]_n - [\beta]_n\}_{n-1} &> \{[\gamma]_n - [\alpha]_n\}_{n-1} + 2d \;, \\
[\gamma]_n - [\beta]_n + 5 \times 10^{-n} &> [\gamma]_n - [\alpha]_n - 5 \times 10^{-n} + 2d \;, \\
[\alpha]_n + 5 \times 10^{-n} &> [\beta]_n - 5 \times 10^{-n} + 2d \;, \\
[\alpha]_n &> [\beta]_n - 10^{-n+1} + 2d
\end{aligned}$$

which proves that $[\alpha]_n > [\beta]_n + d$, i.e. $\alpha > \beta$. ∎

The relations $\alpha \leq \beta$, $\alpha - \gamma \leq \beta - \gamma$, and $\gamma - \beta \leq \gamma - \alpha$ are equivalent.

Proof. The assumption $\alpha - \gamma \leq \beta - \gamma$ implies that $\alpha \leq \beta$, because $\alpha > \beta$ together with $\alpha - \gamma \leq \beta - \gamma$ is absurd. The assumption $\gamma - \beta \leq \gamma - \alpha$ implies that $\alpha - \gamma \leq \beta - \gamma$, because $\gamma - \beta \leq \gamma - \alpha$ together with $\alpha - \gamma > \beta - \gamma$ is absurd. The assumption $\alpha \leq \beta$ implies that $\gamma - \beta \leq \gamma - \alpha$, because $\alpha \leq \beta$ together with $\gamma - \beta > \gamma - \alpha$ is absurd. ∎

2.2.4 Order relations and absolute differences

For all real numbers α, β the relations $\alpha - \beta \leq |\alpha - \beta|$ and $\beta - \alpha \leq |\alpha - \beta|$ hold. On the other hand: $\alpha \leq \beta$ implies that $|\alpha - \beta| \leq \beta - \alpha$.

Proof. For all positive integers n

$$[\alpha - \beta]_n \leq \big|[\alpha - \beta]_n\big| = [|\alpha - \beta|]_n$$

which proves $\alpha - \beta \leq |\alpha - \beta|$, and

$$[|\beta - \alpha|]_n = \big|[\beta - \alpha]_n\big| = \big|[\alpha - \beta]_n\big| = [|\alpha - \beta|]_n$$

which proves $\beta - \alpha \leq |\alpha - \beta|$.

Now assume $\alpha \leq \beta$ i.e. $[\alpha - \beta]_n \leq 10^{-n}$ for all positive integers n to be true. $[|\alpha - \beta|]_n$ coincides either with $[\beta - \alpha]_n$ or $[\alpha - \beta]_n$, depending on which of these two numbers is nonnegative. In the case

$$[|\alpha - \beta|]_n = [\alpha - \beta]_n = -[\beta - \alpha]_n \, ,$$

i.e. $[\beta - \alpha]_n \leq 0$, the inequality $[\beta - \alpha]_n \geq -10^{-n}$ certainly holds and this implies that

$$[|\alpha - \beta|]_n = [\alpha - \beta]_n \leq 10^{-n} \leq [\beta - \alpha]_n + 2 \times 10^{-n} \, .$$

In the case

$$[|\alpha - \beta|]_n = [\beta - \alpha]_n$$

the inequality

$$[|\alpha - \beta|]_n \leq [\beta - \alpha]_n + 2 \times 10^{-n}$$

is evidently true. Using the inequality

$$[|\alpha - \beta|]_n \leq [\beta - \alpha]_n + 2 \times 10^{-n}$$

that follows from $\alpha \leq \beta$ – now with $n + 1$ instead of n – we derive

$$\begin{aligned}
[|\alpha - \beta| - (\beta - \alpha)]_n &\leq [|\alpha - \beta|]_{n+1} - [\beta - \alpha]_{n+1} + 5 \times 10^{-n-1} \\
&\leq [\beta - \alpha]_{n+1} + 2 \times 10^{-n-1} \\
&\quad - [\beta - \alpha]_{n+1} + 5 \times 10^{-n-1} \\
&= 7 \times 10^{-n-1} \leq 10^{-n} \, ,
\end{aligned}$$

which proves $|\alpha - \beta| \leq \beta - \alpha$. ∎

For all real numbers α, β, γ the two relations $\alpha - \beta \leq \gamma$ and $\beta - \alpha \leq \gamma$ imply $|\alpha - \beta| \leq \gamma$.

Proof. Let e be an arbitrary positive decimal number. Then there exists a positive integer j_1 such that for all $n \geq j_1$

$$[\alpha - \beta]_n \leq [\gamma]_n + e,$$

and there exists a positive integer j_2 such that for all $n \geq j_2$

$$[\beta - \alpha]_n \leq [\gamma]_n + e.$$

Defining $j = \max(j_1, j_2)$ and observing that $[|\alpha - \beta|]_n$ coincides either with $[\alpha - \beta]_n$ or with $[\beta - \alpha]_n$, we conclude for all $n \geq j$

$$[|\alpha - \beta|]_n \leq [\gamma]_n + e$$

which proves $|\alpha - \beta| \leq \gamma$. ∎

2.2.5 Triangle inequalities

Triangle inequality – first version. *For any real numbers α, β, γ we have*

$$||\alpha - \gamma| - |\beta - \gamma|| \leq |\alpha - \beta|.$$

Proof. Let e be an arbitrary positive decimal number. Define the integer j so big that $e \geq 2 \times 10^{-j}$. We then conclude for all integers $n \geq j$

$$\begin{aligned}
[||\alpha - \gamma| - |\beta - \gamma||]_n &= \left\{[|\alpha - \gamma|]_{n+1} - [|\beta - \gamma|]_{n+1}\right\}_n \\
&\leq \left|[\alpha - \gamma]_{n+1}\right| - \left|[\beta - \gamma]_{n+1}\right| + 5 \times 10^{-n-1} \\
&= \left|\left\{[\alpha]_{n+2} - [\gamma]_{n+2}\right\}_{n+1}\right| - \left|\left\{[\beta]_{n+2} - [\gamma]_{n+2}\right\}_{n+1}\right| \\
&\quad + 5 \times 10^{-n-1} \\
&\leq \left|[\alpha]_{n+2} - [\gamma]_{n+2}\right| - \left|[\beta]_{n+2} - [\gamma]_{n+2}\right| \\
&\quad + 5 \times 10^{-n-2} + 5 \times 10^{-n-2} + 5 \times 10^{-n-1} \\
&\leq \left|[\alpha]_{n+2} - [\beta]_{n+2}\right| + 6 \times 10^{-n-1} \\
&\leq \left\{|[\alpha]_{n+2} - [\beta]_{n+2}|\right\}_{n+1} + 5 \times 10^{-n-2} \\
&\quad + 6 \times 10^{-n-1} \\
&= [|\alpha - \beta|]_{n+1} + 65 \times 10^{-n-2} \\
&\leq [|\alpha - \beta|]_n + 10^{-n} + 10^{-n-1} + 65 \times 10^{-n-2} \\
&= [|\alpha - \beta|]_n + 175 \times 10^{-n-2} \\
&< [|\alpha - \beta|]_n + 2 \times 10^{-j} \leq [|\alpha - \beta|]_n + e,
\end{aligned}$$

and this proves the assertion. ∎

Triangle inequality – second version. *For any real numbers α, β, γ and any decimal numbers d, e the two inequalities*

$$|\alpha - \beta| \leq d \quad \text{and} \quad |\beta - \gamma| \leq e$$

imply

$$|\alpha - \gamma| \leq d + e.$$

Proof. The relation $|\beta - \gamma| \leq e$ leads to

$$|\alpha - \gamma| - e \leq |\alpha - \gamma| - |\beta - \gamma|,$$

and together with the first version of the triangle inequality to

$$|\alpha - \gamma| - e \leq |\alpha - \beta| \leq d.$$

As the assumption $|\alpha - \gamma| > d + e$ leads to the contradiction

$$|\alpha - \gamma| - e > (d + e) - e = d$$

we in fact have $|\alpha - \gamma| \leq d + e$. ∎

Triangle inequality – third version. *For any real numbers $\alpha_0, \alpha_1, \ldots, \alpha_n$ and any decimal numbers $d_1, \ldots, d_n$ the n inequalities*

$$|\alpha_0 - \alpha_1| \leq d_1, \quad \ldots, \quad |\alpha_{n-1} - \alpha_n| \leq d_n$$

imply

$$|\alpha_0 - \alpha_n| \leq d_1 + \ldots + d_n.$$

Proof. If this assertion is already true for $n - 1$, it only remains to conclude

$$|\alpha_0 - \alpha_n| \leq d_1 + \ldots + d_n$$

from the two inequalities

$$|\alpha_0 - \alpha_{n-1}| \leq d_1 + \ldots + d_{n-1} \quad \text{and} \quad |\alpha_{n-1} - \alpha_n| \leq d_n.$$

This, of course, is clear by the second version of the triangle inequality. ∎

2.2.6 Interpolation and Dichotomy

Approximation lemma. *For any real number α and any positive integer n we have*
$$|\alpha - [\alpha]_n| \leq 10^{-n}.$$

Proof. Let e be an arbitrary positive decimal number and define the integer $j \geq n$ so big that $e \geq 10^{-j}$. We then have for all $m \geq j$

$$\begin{aligned}
\left[|\alpha - [\alpha]_n|\right]_m &= \left|[\alpha - [\alpha]_n]_m\right| = \left|\left\{[\alpha]_{m+1} - [[\alpha]_n]_{m+1}\right\}_m\right| \\
&= \left|\left\{[\alpha]_{m+1} - [\alpha]_n\right\}_m\right| \leq \left|[\alpha]_{m+1} - [\alpha]_n\right| + 5 \times 10^{-m-1} \\
&\leq 10^{-m-1} + 10^{-n} + 5 \times 10^{-m-1} < 10^{-n} + e \\
&= \left[10^{-n}\right]_m + e
\end{aligned}$$

which proves the assertion. ∎

Nesting lemma. *For any real number α and any positive integer n there exists an integer p such that*
$$(p-1) \times 10^{-n} \leq \alpha \leq (p+1) \times 10^{-n}.$$

Proof. As $[\alpha]_n = p \times 10^{-n}$ implies that
$$\left|\alpha - p \times 10^{-n}\right| \leq 10^{-n}$$

by the approximation lemma, thus proving
$$\alpha - p \times 10^{-n} \leq 10^{-n} \quad \text{and} \quad p \times 10^{-n} - \alpha \leq 10^{-n},$$

the rules about order relations and differences lead to the assertion. ∎

Interpolation lemma. *For any real numbers α, β with $\alpha > \beta$ and any positive integer k there exist k decimal numbers $c_1, c_2, \ldots, c_k$ with the property*
$$\alpha > c_1 > c_2 > \ldots > c_k > \beta.$$

Proof. We first start with case $k = 1$. The inequality $\alpha > \beta$ allows us to construct a positive integer n such that
$$[\alpha - \beta]_n > 10^{-n}, \quad \text{i.e.} \quad \left\{[\alpha]_{n+1} - [\beta]_{n+1}\right\}_n > 10^{-n},$$

thus implying
$$[\alpha]_{n+1} - [\beta]_{n+1} > 10^{-n} - 5 \times 10^{-n-1} = 5 \times 10^{-n-1}.$$

As the numbers $[\alpha]_{n+1}$, $[\beta]_{n+1}$, $5 \times 10^{-n-1}$ are decimal numbers with exactly $(n+1)$ decimal places, we even have the inequality

$$[\alpha]_{n+1} - [\beta]_{n+1} \geq 6 \times 10^{-n-1} .$$

We now define $c_1 = c$ to be

$$c = [\beta]_{n+1} + 3 \times 10^{-n-1} .$$

Since c is a decimal number with exactly $(n+1)$ decimal places, we can identify $[c]_{n+2}$ with c itself. Therefore we have

$$\begin{aligned}
{[c - \beta]_{n+1}} &= \{[c]_{n+2} - [\beta]_{n+2}\}_{n+1} \geq [c]_{n+2} - [\beta]_{n+2} - 5 \times 10^{-n-2} \\
&= [\beta]_{n+1} - [\beta]_{n+2} + 3 \times 10^{-n-1} - 5 \times 10^{-n-2} \\
&\geq -10^{-n-2} - 10^{-n-1} + 3 \times 10^{-n-1} - 5 \times 10^{-n-2} \\
&= 14 \times 10^{-n-2} > 10^{-n-1} ,
\end{aligned}$$

i.e. $c > \beta$, and we equally have

$$\begin{aligned}
{[\alpha - c]_{n+1}} &= \{[\alpha]_{n+2} - [c]_{n+2}\}_{n+1} \geq [\alpha]_{n+2} - [c]_{n+2} - 5 \times 10^{-n-2} \\
&= [\alpha]_{n+2} - [\beta]_{n+1} - 3 \times 10^{-n-1} - 5 \times 10^{-n-2} \\
&\geq [\alpha]_{n+1} - [\beta]_{n+1} - 10^{-n-2} - 10^{-n-1} \\
&\quad -3 \times 10^{-n-1} - 5 \times 10^{-n-2} \\
&\geq 6 \times 10^{-n-1} - 46 \times 10^{-n-2} = 14 \times 10^{-n-2} > 10^{-n-1} ,
\end{aligned}$$

i.e. $\alpha > c$.

Now let us assume that we already have $k - 1$ decimal numbers $c_1, \ldots, c_{k-1}$ fulfilling

$$\alpha > c_1 > \ldots > c_{k-1} > \beta .$$

A copy of the argument above, applied to $c_{k-1} > \beta$, proves the existence of a decimal number c_k with $c_{k-1} > c_k > \beta$, and thus proves the interpolation lemma.

∎

Theorem of EUDOXOS and ARCHIMEDES. *To each real number $\varepsilon > 0$ it is possible to construct a positive decimal number $e < \varepsilon$.*

Dichotomy lemma. *For any real numbers α, β, γ the inequality $\alpha > \beta$ implies at least one of the two inequalities $\alpha > \gamma$ or $\gamma > \beta$.*

Proof. The inequality $\alpha > \beta$ allows us to construct a positive integer n with

$$[\alpha - \beta]_n > 10^{-n} , \quad \text{i.e.} \quad \{[\alpha]_{n+1} - [\beta]_{n+1}\}_n > 10^{-n} ,$$

and therefore

$$[\alpha]_{n+1} - [\beta]_{n+1} > 10^{-n} - 5 \times 10^{-n-1} = 5 \times 10^{-n-1} .$$

As the numbers $[\alpha]_{n+1}, [\beta]_{n+1}, 5 \times 10^{-n-1}$ are decimal numbers with exactly $(n+1)$ decimal places, we even have the inequality

$$[\alpha]_{n+1} - [\beta]_{n+1} \geq 6 \times 10^{-n-1}.$$

The number $[\gamma]_{n+1}$ is also a decimal number with exactly $(n+1)$ decimal places. This implies that one of the inequalities

$$[\gamma]_{n+1} \geq [\beta]_{n+1} + 3 \times 10^{-n-1} \quad \text{or} \quad [\gamma]_{n+1} \leq [\beta]_{n+1} + 3 \times 10^{-n-1}$$

has to be true.
In the case

$$[\gamma]_{n+1} \geq [\beta]_{n+1} + 3 \times 10^{-n-1}$$

we derive

$$\begin{aligned}
[\gamma - \beta]_{n+2} &= \left\{[\gamma]_{n+3} - [\beta]_{n+3}\right\}_{n+2} \geq [\gamma]_{n+3} - [\beta]_{n+3} - 5 \times 10^{-n-3} \\
&\geq [\gamma]_{n+1} - 10^{-n-1} - 10^{-n-3} \\
&\quad - [\beta]_{n+1} - 10^{-n-1} - 10^{-n-3} - 5 \times 10^{-n-3} \\
&\geq [\gamma]_{n+1} - [\beta]_{n+1} - 2 \times 10^{-n-1} - 7 \times 10^{-n-3} \\
&\geq 3 \times 10^{-n-1} - 2 \times 10^{-n-1} - 7 \times 10^{-n-3} = 93 \times 10^{-n-3} \\
&> 9 \times 10^{-n-2} > 10^{-n-2},
\end{aligned}$$

i.e. $\gamma > \beta$.
In the case

$$[\gamma]_{n+1} \leq [\beta]_{n+1} + 3 \times 10^{-n-1}$$

we derive

$$\begin{aligned}
[\alpha - \gamma]_{n+2} &= \left\{[\alpha]_{n+3} - [\gamma]_{n+3}\right\}_{n+2} \geq [\alpha]_{n+3} - [\gamma]_{n+3} - 5 \times 10^{-n-3} \\
&\geq [\alpha]_{n+1} - 10^{-n-1} - 10^{-n-3} \\
&\quad - [\gamma]_{n+1} - 10^{-n-1} - 10^{-n-3} - 5 \times 10^{-n-3} \\
&\geq [\alpha]_{n+1} - [\beta]_{n+1} - 3 \times 10^{-n-1} - 2 \times 10^{-n-1} - 7 \times 10^{-n-3} \\
&\geq 6 \times 10^{-n-1} - 5 \times 10^{-n-1} - 7 \times 10^{-n-3} = 93 \times 10^{-n-3} \\
&> 9 \times 10^{-n-2} > 10^{-n-2},
\end{aligned}$$

i.e. $\alpha > \gamma$. ∎

Localization lemma. *For any integer $n > 1$, any real numbers $\alpha_0, \alpha_1, \ldots, \alpha_n$ with $\alpha_0 < \alpha_1 < \ldots < \alpha_n$, and any real number β with $\alpha_0 < \beta < \alpha_n$ there exists a positive integer $m < n$ such that $\alpha_{m-1} < \beta < \alpha_{m+1}$.*

Proof. The dichotomy lemma implies from $\alpha_1 < \alpha_2$ that one of the two inequalities $\beta < \alpha_2$ or $\alpha_1 < \beta$ must hold. In the case $\beta < \alpha_2$ (which is for sure if $n = 2$)

we define $m = 1$ and thereby finish the proof. In the case $n > 2$ and $\alpha_1 < \beta$ we transfer the argument to the sequence $\alpha_1, \ldots, \alpha_n$ and the inequality $\alpha_2 < \alpha_3$. This procedure goes on, till we reach the sequence $\alpha_{m-1}, \ldots, \alpha_n$ and the inequality $\alpha_m < \alpha_{m+1}$ where the first of the two possible cases $\beta < \alpha_{m+1}$ or $\alpha_m < \beta$ in fact holds – a situation that at least at the stage $m = n - 1$ must happen. ∎

2.3 Equality and apartness

2.3.1 Definition and criteria

The *apartness* $\alpha \neq \beta$ signifies for real numbers α, β that one can find an integer $n > 0$ with the property $|[\alpha - \beta]_n| > 10^{-n}$. The *equality* $\alpha = \beta$ signifies for real numbers α, β that one can prove $|[\alpha - \beta]_n| \leq 10^{-n}$ for all integers $n > 0$. Otherwise stated: Two real numbers α, β are apart if and only if one of the two inequalities $\alpha > \beta$ or $\beta > \alpha$ holds. They are identical if and only if both of the two inequalities $\alpha \leq \beta$ and $\beta \leq \alpha$ hold.

The indirect proof. *A contradiction of the assumption $\alpha \neq \beta$ proves the equality $\alpha = \beta$.*

Proof. This follows immediately from the indirect proof in the preceding section. ∎

We again strictly *avoid* to derive $\alpha \neq \beta$ from the fact that $\alpha = \beta$ leads to a contradiction for the following reason: The impossibility of $|[\alpha - \beta]_n| \leq 10^{-n}$ for all integers $n > 0$ in general *does not* show how to pick up an integer $n > 0$ that guarantees $|[\alpha - \beta]_n| > 10^{-n}$.

Criterion of apartness. *The apartness $\alpha \neq \beta$ holds for real numbers α, β if and only if one can find a real number $\delta > 0$ and a positive integer j such that for all integers $n \geq j$ the inequality $|[\alpha]_n - [\beta]_n| > \delta$ follows.*

Proof. On the one hand, suppose that $\alpha \neq \beta$: In the case $\alpha > \beta$, there exists a positive decimal number d and a positive integer j such that $n \geq j$ implies that $[\alpha]_n - [\beta]_n > d$. We now set $\delta = d$ and gain

$$\left|[\alpha]_n - [\beta]_n\right| \geq [\alpha]_n - [\beta]_n > \delta$$

for all $n \geq j$. In the case $\beta > \alpha$, there exists a positive decimal number d and a positive integer j such that $n \geq j$ implies that $[\beta]_n - [\alpha]_n > d$. We again set $\delta = d$ with the same result

$$\left|[\alpha]_n - [\beta]_n\right| \geq [\beta]_n - [\alpha]_n > \delta$$

for all $n \geq j$.

If on the other hand, the existence of a positive real δ and a positive integer j with the property that $n \geq j$ implies that $\left|[\alpha]_n - [\beta]_n\right| > \delta$ is supposed, then the interpolation lemma proves the existence of a decimal number d such that

$$\left|[\alpha]_n - [\beta]_n\right| > \delta > d > 0$$

for all $n \geq j$. We now define the integer $k \geq j$ so big that $d \geq 2 \times 10^{-k}$. It is clear that for all $n \geq k$ resp. for all $m \geq k$ either $[\alpha]_n > [\beta]_n + d$ or $[\beta]_n > [\alpha]_n + d$ resp. either $[\alpha]_m > [\beta]_m + d$ or $[\beta]_m > [\alpha]_m + d$. If $[\alpha]_n > [\beta]_n + d$ and $[\beta]_m > [\alpha]_m + d$ would be the case, the chain of inequalities

$$[\alpha]_n > [\beta]_n + d \geq [\beta]_m + d - 10^{-n} - 10^{-m} > [\alpha]_m + 2d - 10^{-n} - 10^{-m}$$

would lead to

$$\begin{aligned}[\alpha]_n - [\alpha]_m &> 2d - 10^{-n} - 10^{-m} \geq 4 \times 10^{-k} - 2 \times 10^{-k} \\ &= 2 \times 10^{-k} \geq 10^{-n} + 10^{-m}\end{aligned}$$

in contradiction to the fact that α is a real number. This proves that for all $n \geq k$ and all $m \geq k$ the inequality $[\alpha]_m > [\beta]_m + d$ follows from $[\alpha]_n > [\beta]_n + d$. The same argument shows that for all $n \geq k$ and all $m \geq k$ the inequality $[\beta]_m > [\alpha]_m + d$ follows from $[\beta]_n > [\alpha]_n + d$. Therefore we have either $\alpha > \beta$ or $\alpha < \beta$. In any case: $\alpha \neq \beta$ is correct. ∎

Criterion of equality. The equality $\alpha = \beta$ holds for real numbers α, β if and only if for each real number $\varepsilon > 0$ there exists a positive integer j such that for all integers $n \geq j$ the inequality $|[\alpha]_n - [\beta]_n| < \varepsilon$ follows.

Proof. On the one hand, suppose that $\alpha = \beta$, i.e. both inequalities $\alpha \leq \beta$ and $\beta \leq \alpha$ hold. For each real $\varepsilon > 0$ the interpolation lemma allows us to construct a decimal number e with $\varepsilon > e > 0$. Further there exists a positive integer j_1 such that $n \geq j_1$ implies that $[\alpha]_n - [\beta]_n \leq e$ and there exists a positive integer j_2 such that $n \geq j_2$ implies that $[\beta]_n - [\alpha]_n \leq e$. We now set $j = \max(j_1, j_2)$ and conclude from the two above inequalities for all $n \geq j$ the inequality $|[\alpha]_n - [\beta]_n| \leq e < \varepsilon$ as asserted.
If, on the other hand, for each real number $\varepsilon > 0$ there exists a positive integer j such that for all integers $n \geq j$ the inequality $|[\alpha]_n - [\beta]_n| < \varepsilon$ follows, then this assumption a fortiori is true for each positive decimal number e instead of ε, and the fact that $n \geq j$ implies that $|[\alpha]_n - [\beta]_n| < e$ proves both $[\alpha]_n \leq [\beta]_n + e$, i.e. $\alpha \leq \beta$, as well as $[\beta]_n \leq [\alpha]_n + e$, i.e. $\beta \leq \alpha$. Therefore we have $\alpha = \beta$. ∎

The difference between the criteria of strong and weak order on the one hand and the criteria of apartness and equality on the other hand is that now we used positive real numbers δ and ε instead of positive decimal numbers d and e. The interpolation lemma guarantees that this exchange is legitimate in general. Therefore the former two criteria can also be formulated in the following way:

Criterion of strict order. *The strict order $\alpha > \beta$ holds for real numbers α, β if and only if one can find a real number $\delta > 0$ and a positive integer j such that for all integers $n \geq j$ the inequality $[\alpha]_n - [\beta]_n > \delta$ follows.*

Criterion of weak order. *The weak order $\alpha \leq \beta$ holds for real numbers α, β if and only if for all real numbers $\varepsilon > 0$ one can find a positive integer j such that for all integers $n \geq j$ the inequality $[\alpha]_n - [\beta]_n < \varepsilon$ follows.*

2.3.2 Properties of equality and apartness

For all real numbers α, β, γ:
1. $\alpha \neq \alpha$ is absurd; we always have $\alpha = \alpha$.
2. $\alpha = \beta$ and $\gamma = \beta$ implies $\alpha = \gamma$.
3. If $\alpha \neq \beta$, then we have either $\alpha \neq \gamma$ or $\beta \neq \gamma$ or both.
4. If $\alpha = \beta$, then $\beta > \gamma$ follows from $\alpha > \gamma$ and $\gamma > \beta$ follows from $\gamma > \alpha$.
5. If $\alpha = \beta$, then $\gamma \leq \beta$ follows from $\gamma \leq \alpha$ and $\beta \leq \gamma$ follows from $\alpha \leq \gamma$.

Proof. 1. This follows from the fact that $\alpha > \alpha$ is absurd.
2. $\alpha \leq \beta$ and $\beta \leq \gamma$ imply $\alpha \leq \gamma$, as well as $\gamma \leq \beta$ and $\beta \leq \alpha$ imply $\gamma \leq \alpha$.
3. $\alpha \neq \beta$ implies either that $\alpha > \beta$ with the consequence $\gamma > \beta$ or $\alpha > \gamma$ and therefore $\alpha \neq \gamma$ or $\beta \neq \gamma$ or both by the dichotomy lemma. Or $\alpha \neq \beta$ implies that $\beta > \alpha$ with the consequence $\gamma > \alpha$ or $\beta > \gamma$ and therefore $\alpha \neq \gamma$ or $\beta \neq \gamma$ or both by the dichotomy lemma.
4. As $\beta \geq \alpha$ and $\alpha > \gamma$ imply $\beta > \gamma$, the first assertion is proved. As $\alpha \geq \beta$ and $\gamma > \alpha$ imply $\gamma > \beta$, the second assertion is proved.
5. As $\alpha \leq \beta$ and $\gamma \leq \alpha$ imply $\gamma \leq \beta$, the first assertion is proved. As $\beta \leq \alpha$ and $\alpha \leq \gamma$ imply $\beta \leq \gamma$, the second assertion is proved. ∎

For all real numbers α, β, γ:
1. If $\alpha = \beta$, then we have $\alpha - \gamma = \beta - \gamma$ and $\gamma - \alpha = \gamma - \beta$.
2. If $\alpha \neq \beta$, then we have $\alpha - \gamma \neq \beta - \gamma$ and $\gamma - \alpha \neq \gamma - \beta$.
3. $|\alpha - \beta| = |\beta - \alpha|$.
4. If $\alpha \leq \beta$, then $|\alpha - \beta| = \beta - \alpha$, if $\beta \leq \alpha$, then $|\alpha - \beta| = \alpha - \beta$.
5. If $\alpha = \beta$, then we have $|\alpha - \gamma| = |\beta - \gamma|$ and $|\gamma - \alpha| = |\gamma - \beta|$.

Proof. 1. $\alpha = \beta$ implies that $\alpha \leq \beta$ with the consequences $\alpha - \gamma \leq \beta - \gamma$ and $\gamma - \beta \leq \gamma - \alpha$. $\alpha = \beta$ also implies that $\beta \leq \alpha$ with the consequences $\beta - \gamma \leq \alpha - \gamma$ and $\gamma - \alpha \leq \gamma - \beta$.
2. We either have $\alpha > \beta$ with the consequences $\alpha - \gamma > \beta - \gamma$ and $\gamma - \beta > \gamma - \alpha$, i.e. $\alpha - \gamma \neq \beta - \gamma$ as well as $\gamma - \beta \neq \gamma - \alpha$. Or we have $\beta > \alpha$ with the consequences $\beta - \gamma > \alpha - \gamma$ and $\gamma - \alpha > \gamma - \beta$, i.e. $\beta - \gamma \neq \alpha - \gamma$ as well as $\gamma - \alpha \neq \gamma - \beta$.
3. As we have $\alpha - \beta \leq |\alpha - \beta|$ and $\beta - \alpha \leq |\alpha - \beta|$, as well as $\beta - \alpha \leq |\beta - \alpha|$ and $\alpha - \beta \leq |\beta - \alpha|$, we may conclude

$$|\alpha - \beta| \leq |\beta - \alpha| \quad \text{as well as} \quad |\beta - \alpha| \leq |\alpha - \beta|.$$

4. The first assertion follows from

$$|\alpha - \beta| \leq \beta - \alpha \leq |\alpha - \beta|,$$

the second from

$$|\alpha - \beta| \leq \alpha - \beta \leq |\alpha - \beta|.$$

5. $\alpha = \beta$ implies that $\alpha \leq \beta$ and $\beta \leq \alpha$ with the consequences

$$\alpha - \gamma \leq \beta - \gamma \leq |\beta - \gamma| \quad \text{and} \quad \gamma - \alpha \leq \gamma - \beta \leq |\beta - \gamma|.$$

Thus we have $|\alpha - \gamma| \le |\beta - \gamma|$. In the same way we derive the consequences

$$\beta - \gamma \le \alpha - \gamma \le |\alpha - \gamma| \quad \text{and} \quad \gamma - \beta \le \gamma - \alpha \le |\alpha - \gamma|.$$

Thus we have $|\beta - \gamma| \le |\alpha - \gamma|$. This proves the equality $|\alpha - \gamma| = |\beta - \gamma|$. The second equality is a consequence of

$$|\alpha - \gamma| = |\gamma - \alpha|, \quad |\beta - \gamma| = |\gamma - \beta|,$$

and point 2 of the former corollary. ∎

Point 4 of this corollary *must not* lead to the assumption that it would be possible to decide between the possibilities $\alpha \le \beta$ or $\beta \le \alpha$ for any real numbers α, β. Quite to the contrary: Point 4 only evaluates $|\alpha - \beta|$ in that cases where we *presuppose* that such a decision is possible.

The inequality of decimal numbers within the system of decimal numbers is equivalent with the apartness of these numbers within the system of the continuum. The equality of decimal numbers within the system of decimal numbers is equivalent with the equality of these numbers within the system of the continuum. The difference (resp. the absolute difference) of decimal numbers within the system of decimal numbers is equal to the difference (resp. the absolute difference) of these numbers within the system of the continuum.

Proof. This follows from the fact that the strong as well as the weak order within the system of decimal numbers is faithfully incorporated within the system of the continuum and from the fact that we already have proved $a-b \le a+(-b) \le a-b$ and $|a - b| \le |a + (-b)| \le |a - b|$, i.e.

$$a + (-b) = a - b \quad \text{and} \quad |a + (-b)| = |a - b|$$

for arbitrary decimal numbers a, b. ∎

2.4 Convergent sequences of real numbers

2.4.1 The limit of convergent sequences

An infinite sequence $A = (a_1, a_2, \ldots, a_n, \ldots)$ of real numbers $a_1, a_2, \ldots, a_n, \ldots$ is called *convergent* if and only if it is possible to construct a real number

$$a = \lim A,$$

the so-called *limit* of the sequence A, with the following property: for any $\varepsilon > 0$ there exists a positive integer j such that for all $n \ge j$ the inequality $|a - a_n| < \varepsilon$ holds.

The capital letters $A, B, C, \ldots$ are used for sequences.

The notation $(\alpha_1, \alpha_2, \ldots, \alpha_n)$ indicates that the sequence consists of a finite number of elements, and the notation $(\alpha_1, \alpha_2, \ldots, \alpha_n, \ldots)$ indicates that the sequence consists of an infinite number of elements.

If A is a convergent infinite sequence, then the relations

$$\alpha' = \lim A \quad \text{and} \quad \alpha'' = \lim A$$

imply the equality $\alpha' = \alpha''$.

Proof. Assume $\alpha' \ne \alpha''$. In other words: $\varepsilon = |\alpha' - \alpha''| > 0$. The interpolation lemma would then allow to construct a decimal number e with $\varepsilon > e > 0$. From $\alpha' = \lim A$ we derive the existence of a positive integer j_1 such that for all $n \ge j_1$ the inequality $|\alpha' - \alpha_n| < e/2$ holds. From $\alpha'' = \lim A$ we derive the existence of a positive integer j_2 such that for all $n \ge j_2$ the inequality $|\alpha'' - \alpha_n| < e/2$ holds. This leads for $n \ge \max(j_1, j_2)$ to the contradiction

$$|\alpha' - \alpha''| \le \frac{e}{2} + \frac{e}{2} = e < \varepsilon = |\alpha' - \alpha''|,$$

i.e. the assumption $\alpha' \ne \alpha''$ turns out to be absurd. ∎

For any real number α the infinite sequence

$$([\alpha]_1, [\alpha]_2, \ldots, [\alpha]_n, \ldots)$$

is convergent with limit α.

Proof. Let $\varepsilon > 0$ be an arbitrary real number. The interpolation lemma allows us to construct a positive decimal number $e < \varepsilon$. We define the positive integer j to be so big that $10^j e \ge 1$. Then for all $n \ge j$, by the approximation lemma,

$$|\alpha - [\alpha]_n| \le 10^{-n} \le 10^{-j} \le e < \varepsilon.$$

∎

2.4.2 Limit and order

Estimation of the limit. Let $A = (\alpha_1, \alpha_2, \ldots, \alpha_n, \ldots)$ denote a convergent sequence and assume the existence of real numbers β, γ and of a positive integer m such that $n \ge m$ implies that $\alpha_n \le \beta$ resp. $\alpha_n \ge \gamma$. Then we have $\lim A \le \beta$ resp. $\lim A \ge \gamma$.

Proof. Assume that $n \ge m$ implies that $\alpha_n \le \beta$ but $\lim A = \alpha > \beta$. By the interpolation lemma there exists a positive decimal number $\varepsilon < \alpha - \beta$, and the convergence of A to α allows us to construct a positive integer j such that $n \ge j$ implies that $|\alpha - \alpha_n| < \varepsilon$, and a fortiori $\alpha - \alpha_n < \varepsilon$. For $n \ge \max(j, m)$ this would lead to the contradiction

$$\alpha - \beta \le \alpha - \alpha_n < \varepsilon < \alpha - \beta.$$

Assume that $n \geq m$ implies that $\alpha_n \geq \gamma$ but $\lim A = \alpha < \gamma$. By the interpolation lemma there exists a positive decimal number $\varepsilon < \gamma - \alpha$, and the convergence of A to α allows us to construct a positive integer j such that $n \geq j$ implies that $|\alpha - \alpha_n| < \varepsilon$, and we have, a fortiori, $\alpha_n - \alpha < \varepsilon$. For $n \geq \max(j, m)$ this would lead to the contradiction

$$\gamma - \alpha \leq \alpha_n - \alpha < \varepsilon < \gamma - \alpha .$$

∎

Let $A = (\alpha_1, \alpha_2, \ldots, \alpha_n, \ldots)$, $B = (\beta_1, \beta_2, \ldots, \beta_n, \ldots)$ denote two sequences. We write $A \leq B$ if for all positive integers n the relations $\alpha_n \leq \beta_n$ hold.

Principle of permanence. *Let*

$$A = (\alpha_1, \alpha_2, \ldots, \alpha_n, \ldots) , \quad B = (\beta_1, \beta_2, \ldots, \beta_n, \ldots)$$

denote two convergent sequences and assume $A \leq B$. Then we have

$$\lim A \leq \lim B .$$

Proof. Define $\beta = \lim B$ and for any positive integer k the decimal number

$$b_k = [\beta]_k + 2 \times 10^{-k} .$$

We set $\varepsilon = 10^{-k}$ and construct a positive integer m_k such that $n \geq m_k$ implies that $|\beta_n - \beta| < 10^{-k}$. The approximation lemma $|\beta - [\beta]_k| \leq 10^{-k}$ on one hand and the triangle inequality on the other hand guarantee for $n \geq m_k$ therefore $\alpha_n \leq \beta_n \leq b_k$. By the estimation of the limit we thus have $\alpha = \lim A \leq b_k$. Furthermore the inequalities

$$|\beta - [\beta]_k| \leq 10^{-k} \quad \text{and} \quad |[\beta]_k - b_k| \leq 2 \times 10^{-k}$$

prove that the sequence $B' = (b_1, b_2, \ldots, b_n, \ldots)$ is convergent with β as limit: For let $\varepsilon > 0$ denote an arbitrary real number and construct by the interpolation lemma a positive decimal number $e < \varepsilon$, then there exists a positive integer j such that $e \geq 3 \times 10^{-j}$. From that we conclude for all $n \geq j$

$$|\beta - b_n| \leq 10^{-n} + 2 \times 10^{-n} \leq 3 \times 10^{-j} \leq e < \varepsilon .$$

For all positive integers n the inequality $\alpha \leq b_n$ holds and therefore – again we refer to the estimation of the limit – we have in fact $\alpha \leq \lim B' = \beta$. ∎

2.4.3 Limit and differences

Let $A = (\alpha_1, \alpha_2, \ldots, \alpha_n, \ldots)$, $B = (\beta_1, \beta_2, \ldots, \beta_n, \ldots)$ denote two sequences. We write $A - B$ for the sequence consisting of the differences $\alpha_1 - \beta_1, \alpha_2 - \beta_2, \ldots, \alpha_n - \beta_n, \ldots$ and write $|A - B|$ for the sequence consisting of the absolute differences $|\alpha_1 - \beta_1|, |\alpha_2 - \beta_2|, \ldots, |\alpha_n - \beta_n|, \ldots$.

Let $A = (\alpha_1, \alpha_2, \ldots, \alpha_n, \ldots)$, $B = (\beta_1, \beta_2, \ldots, \beta_n, \ldots)$ denote two convergent sequences. Then $A - B$ is also a convergent sequence, and we have

$$\lim (A - B) = \lim A - \lim B.$$

Proof. Define $\alpha = \lim A$, $\beta = \lim B$, and let ε be an arbitrary positive real number. The interpolation lemma allows us to construct a positive decimal number $e < \varepsilon$. There exist two positive integers j_1, j_2 such that $n \geq j_1$ implies that $|\alpha - \alpha_n| < e/4$, and such that $n \geq j_2$ implies that $|\beta - \beta_n| < e/4$. We now set $j = \max(j_1, j_2)$ and define the integer m to be so big that $68 \times 10^{-m-1} \leq e$. First, the triangle inequality shows for all $n \geq j$ that the three inequalities

$$|[\alpha]_{m+1} - \alpha| \leq 10^{-m-1}$$
$$|\alpha - \alpha_n| \leq \frac{e}{4}$$
$$|\alpha_n - [\alpha_n]_{m+1}| \leq 10^{-m-1}$$

yield

$$|[\alpha]_{m+1} - [\alpha_n]_{m+1}| \leq \frac{e}{4} + 2 \times 10^{-m-1}.$$

Equally the triangle inequality shows for all $n \geq j$ that the three inequalities

$$|[\beta]_{m+1} - \beta| \leq 10^{-m-1}$$
$$|\beta - \beta_n| \leq \frac{e}{4}$$
$$|\beta_n - [\beta_n]_{m+1}| \leq 10^{-m-1}$$

yield

$$|[\beta]_{m+1} - [\beta_n]_{m+1}| \leq \frac{e}{4} + 2 \times 10^{-m-1}.$$

Second, we use the following estimation in which only decimal numbers are involved:

$$\left|([\alpha]_{m+1} - [\beta]_{m+1}) - ([\alpha_n]_{m+1} - [\beta_n]_{m+1})\right|$$
$$= \left|[\alpha]_{m+1} - [\beta]_{m+1} - [\alpha_n]_{m+1} + [\beta_n]_{m+1}\right|$$
$$\leq \left|[\alpha]_{m+1} - [\alpha_n]_{m+1}\right| + \left|[\beta_n]_{m+1} - [\beta]_{m+1}\right|$$
$$\leq 2\left(\frac{e}{4} + 2 \times 10^{-m-1}\right) = \frac{e}{2} + 4 \times 10^{-m-1}.$$

Third the approximation lemma and the rounding of decimal numbers, applied to $\alpha - \beta$, i.e.

$$|(\alpha - \beta) - [\alpha - \beta]_m| \leq 10^{-m}$$
$$|[\alpha - \beta]_m - ([\alpha]_{m+1} - [\beta]_{m+1})| \leq 5 \times 10^{-m-1}$$

imply by the triangle inequality

$$|(\alpha - \beta) - ([\alpha]_{m+1} - [\beta]_{m+1})| \leq 15 \times 10^{-m-1} .$$

Equally the approximation lemma and the rounding of decimal numbers, applied to $\alpha_n - \beta_n$, i.e.

$$|(\alpha_n - \beta_n) - [\alpha_n - \beta_n]_m| \leq 10^{-m}$$
$$\left|[\alpha_n - \beta_n]_m - \left([\alpha_n]_{m+1} - [\beta_n]_{m+1}\right)\right| \leq 5 \times 10^{-m-1}$$

imply by the triangle inequality

$$\left|(\alpha_n - \beta_n) - \left([\alpha_n]_{m+1} - [\beta_n]_{m+1}\right)\right| \leq 15 \times 10^{-m-1} .$$

So we have the three inequalities

$$|(\alpha - \beta) - ([\alpha]_{m+1} - [\beta]_{m+1})| \leq 15 \times 10^{-m-1}$$
$$\left|([\alpha]_{m+1} - [\beta]_{m+1}) - \left([\alpha_n]_{m+1} - [\beta_n]_{m+1}\right)\right| \leq \frac{e}{2} + 4 \times 10^{-m-1}$$
$$\left|\left([\alpha_n]_{m+1} - [\beta_n]_{m+1}\right) - (\alpha_n - \beta_n)\right| \leq 15 \times 10^{-m-1}$$

which the triangle inequality combines to

$$|(\alpha - \beta) - (\alpha_n - \beta_n)| \leq 15 \times 10^{-m-1} + \left(\frac{e}{2} + 4 \times 10^{-m-1}\right)$$
$$+ 15 \times 10^{-m-1}$$
$$= \frac{e}{2} + 34 \times 10^{-m-1} \leq e < \varepsilon .$$

∎

Let $A = (\alpha_1, \alpha_2, \ldots, \alpha_n, \ldots)$, $B = (\beta_1, \beta_2, \ldots, \beta_n, \ldots)$ *denote two convergent sequences. Then* $|A - B|$ *is also a convergent sequence, and we have*

$$\lim |A - B| = |\lim A - \lim B| .$$

Proof. Define $\alpha = \lim A$, $\beta = \lim B$, and let ε be an arbitrary positive real number. The interpolation lemma allows us to construct a positive decimal number $e < \varepsilon$. There exist two positive integers j_1, j_2 such that $n \geq j_1$ implies that $|\alpha - \alpha_n| < e/2$, and such that $n \geq j_2$ implies that $|\beta - \beta_n| < e/2$. We now set

46 2. Real numbers

$j = \max(j_1, j_2)$. From the triangle inequality we have for all $n \geq j$ on the one hand

$$|\alpha - \beta| - |\alpha_n - \beta| \leq |\alpha - \alpha_n| < \frac{e}{2},$$
$$|\alpha_n - \beta| - |\alpha - \beta| \leq |\alpha_n - \alpha| < \frac{e}{2},$$

on the other hand

$$|\alpha_n - \beta| - |\alpha_n - \beta_n| \leq |\beta - \beta_n| < \frac{e}{2},$$
$$|\alpha_n - \beta_n| - |\alpha_n - \beta| \leq |\beta_n - \beta| < \frac{e}{2},$$

therefore

$$||\alpha - \beta| - |\alpha_n - \beta|| \leq |\alpha - \alpha_n| < \frac{e}{2},$$
$$||\alpha_n - \beta| - |\alpha_n - \beta_n|| \leq |\beta - \beta_n| < \frac{e}{2},$$

which proves again by the triangle inequality

$$||\alpha - \beta| - |\alpha_n - \beta_n|| \leq \frac{e}{2} + \frac{e}{2} < \varepsilon.$$

∎

2.4.4 The convergence criterion

CAUCHY-**criterion.** *Let $A = (\alpha_1, \alpha_2, \ldots, \alpha_n, \ldots)$ denote an infinite sequence of real numbers. A is convergent if and only if it is possible to construct for any real $\varepsilon > 0$ a positive integer k such that $n \geq k$ and $m \geq k$ imply $|\alpha_m - \alpha_n| < \varepsilon$.*

Proof. First assume A to be convergent with limit $\alpha = \lim A$. Let $\varepsilon > 0$ denote an arbitrary real number. The interpolation lemma allows us to construct a positive decimal number $e < \varepsilon$. The convergence of A allows us to construct a positive integer k such that $n \geq k$ implies that $|\alpha - \alpha_n| < e/2$. Equally $|\alpha_m - \alpha| < e/2$ follows from $m \geq k$. The triangle inequality therefore proves for all $n \geq k, m \geq k$

$$|\alpha_m - \alpha_n| \leq \frac{e}{2} + \frac{e}{2} = e < \varepsilon,$$

the condition of the CAUCHY-criterion.

Next, assume that the condition of the CAUCHY-criterion is fulfilled. Let j denote an arbitrary positive integer and set $\varepsilon = 10^{-j-1}$. Then it is possible to construct a positive integer k_j such that $n \geq k_j$ and $m \geq k_j$ imply

$$|\alpha_m - \alpha_n| < 10^{-j-1}.$$

We now define

$$[\alpha]_j = \left\{[\alpha_{k_j}]_{j+1}\right\}_j,$$

and immediately observe that $[\alpha]_j$ is a decimal number with exactly j decimal places. Further we prove that the sequence of these decimal numbers $[\alpha]_1, [\alpha]_2,$ ..., $[\alpha]_n, \ldots$ in fact define a real number α. To do this it is just enough to look at the five inequalities

$$\left|[\alpha]_n - [\alpha_{k_n}]_{n+1}\right| \leq 5 \times 10^{-n-1}$$
$$\left|[\alpha_{k_n}]_{n+1} - \alpha_{k_n}\right| \leq 10^{-n-1}$$
$$\left|\alpha_{k_n} - \alpha_{k_m}\right| \leq \max\left(10^{-n-1}, 10^{-m-1}\right) \leq 10^{-n-1} + 10^{-m-1}$$
$$\left|\alpha_{k_m} - [\alpha_{k_m}]_{m+1}\right| \leq 10^{-m-1}$$
$$\left|[\alpha_{k_m}]_{m+1} - [\alpha]_m\right| \leq 5 \times 10^{-m-1}$$

that hold for all positive integers n, m. The first and the fifth follow from the rounding of decimal numbers, the second and the fourth follow from the approximation lemma, and the third follows from the definition of k_n resp. k_m. Using the triangle inequality we obtain the result

$$\begin{aligned}|[\alpha]_n - [\alpha]_m| &\leq 5 \times 10^{-n-1} + 10^{-n-1} + \left(10^{-n-1} + 10^{-m-1}\right) \\ &\quad + 10^{-m-1} + 5 \times 10^{-m-1} \\ &= 7 \times 10^{-n-1} + 7 \times 10^{-m-1} \leq 10^{-n} + 10^{-m}.\end{aligned}$$

Finally, we show that A is convergent with α as limit. Let $\varepsilon > 0$ be an arbitrary real number, and construct by the interpolation lemma a positive decimal number $e < \varepsilon$. Now define the positive integer j to be so big that $e \geq 17 \times 10^{-j-1}$. For all $n \geq k_j$ we have the four inequalities

$$\left|\alpha_n - \alpha_{k_j}\right| \leq 10^{-j-1},$$
$$\left|\alpha_{k_j} - [\alpha_{k_j}]_{j+1}\right| \leq 10^{-j-1},$$
$$\left|[\alpha_{k_j}]_{j+1} - [\alpha]_j\right| \leq 5 \times 10^{-j-1},$$
$$\left|[\alpha]_j - \alpha\right| \leq 10^{-j}$$

The first follows from the definition of k_j and from $n \geq k_j$, the second and the fourth follow from the approximation lemma, and the third from the rounding of decimal numbers. Again by the triangle inequality these lead to

$$\begin{aligned}|\alpha_n - \alpha| &\leq 10^{-j-1} + 10^{-j-1} + 5 \times 10^{-j-1} + 10^{-j} \\ &= 17 \times 10^{-j-1} \leq e < \varepsilon.\end{aligned}$$

This in fact proves $\alpha = \lim A$. ∎

We now, in the following chapter, define a far more general frame that also allows us to formulate a CAUCHY-criterion and that confirms the importance of this result.

3
Metric spaces

3.1 Metric spaces and complete metric spaces

3.1.1 Definition of metric spaces

A system S of objects $x, y, z, \ldots$ that are called *points* is a *metric space* if and only if it is possible to construct out of any pair x, y of two points a real number $\|x - y\|$ that is called the *metric* or the *distance* of these points x and y. The notation $\|x - y\|$ obviously generalizes the absolute difference $|\alpha - \beta|$ of two real numbers α, β. Nevertheless, one has to observe that in metric spaces the existence of a "difference" $x - y$ of points x, y itself in general is not presupposed.

For the metric the following three properties must hold:
The metric is positive: we always have $\|x - y\| \geq 0$.
The metric is definite: we have $\|x - y\| > 0$ if and only if $x \neq y$.
The metric obeys the triangle inequality: we always have

$$\|z - x\| - \|z - y\| \leq \|x - y\| .$$

We presuppose that an *apartness* and an *equality* relation are defined within the metric space S. These two relations obey the following rules:
$x \neq y$ together with $x = y$ is absurd.
If $x \neq y$ is contradictory, then $x = y$.
$x \neq x$ is absurd; we therefore always have $x = x$.
$x = y$ and $z = y$ implies that $x = z$.
$x \neq y$ implies, for any point z, either that $x \neq z$ or that $y \neq z$ or both.
$x = y$ implies $\|x - z\| = \|y - z\|$ and $\|z - x\| = \|z - y\|$.

The letters x, y, z are used for points, the letters S, T, U, V, W are used for metric spaces. If all points of the metric space T coincide with points of the metric space S, then T is called a *subspace* or a *subset* of S; we write $T \subseteq S$. If it is clear from the context that S is the metric space under discussion, T is simply called a *set*. We write $x \in T$ to express that x is a point of T. The letters X, Y, Z usually denote *sequences* of points of a metric space. We write $x \in X$ to express that we can find a positive integer n such that $x = x_n$ is the n-th element of X. *Sequences of metric spaces* are denoted by Greek capital letters like Σ, Ξ.

Symmetry of the metric. *The metric is symmetric: we always have*

$$\|x - y\| = \|y - x\| .$$

Proof. This follows from $\alpha - 0 = \alpha$, a formula that holds for any real α, and from $\|x - x\| = 0$: Taking $z = y$ in the triangle inequality

$$\|z - x\| - \|z - y\| \leq \|x - y\|$$

one gets

$$\|y - x\| \leq \|x - y\| .$$

Taking $z = x$ in the triangle inequality

$$\|z - y\| - \|z - x\| \leq \|y - x\|$$

one gets

$$\|x - y\| \leq \|y - x\| .$$

■

The symmetry of the metric allows us to formulate the triangle inequality as

$$\|x - z\| - \|y - z\| \leq \|x - y\| \quad \text{or} \quad \|x - z\| - \|z - y\| \leq \|x - y\| .$$

Triangle inequality – second version. *For arbitrary points x, y, z, and for arbitrary decimal numbers d', d'', the inequality*

$$\|x - z\| \leq d' + d''$$

follows from $\|x - y\| \leq d'$ and $\|y - z\| \leq d''$.

Proof. Assume $\|y - z\| \leq d''$. Then we have

$$\|x - z\| - d'' \leq \|x - z\| - \|y - z\|$$

and by the triangle inequality

$$\|x - z\| - d'' \leq \|x - y\| \leq d' .$$

This immediately leads to

$$\|x - z\| - d'' \leq \|x - y\| \leq d' .$$

∎

Triangle inequality – third version. *For arbitrary $n + 1$ points $x_0, x_1, x_2, \ldots, x_n$, and for arbitrary n decimal numbers $d_1, d_2, \ldots, d_n$, the inequality*

$$\|x_0 - x_n\| \leq d_1 + d_2 + \ldots + d_n$$

follows from

$$\|x_0 - x_1\| \leq d_1 , \quad \|x_1 - x_2\| \leq d_2 , \quad \ldots , \quad \|x_{n-1} - x_n\| \leq d_n .$$

Proof. This, of course, follows from the second version by induction. ∎

3.1.2 Fundamental sequences

A sequence $X = (x_1, x_2, \ldots, x_n, \ldots)$ of points $x_0, x_1, x_2, \ldots, x_n, \ldots$ is called a *fundamental sequence* if and only if it is possible to construct, for any real $\varepsilon > 0$, a positive integer j such that, for all $m \geq j$ and all $n \geq j$, the inequality $\|x_n - x_m\| < \varepsilon$ holds.

Let $X = (x_1, x_2, \ldots, x_n, \ldots)$ and $Y = (y_1, y_2, \ldots, y_n, \ldots)$ denote fundamental sequences. Then the sequence $\|X - Y\|$ consisting of the real numbers $\|x_1 - y_1\|$, $\|x_2 - y_2\|, \ldots, \|x_n - y_n\|, \ldots$ is convergent.

Proof. Let $\varepsilon > 0$ denote an arbitrary real number. By the interpolation lemma, there exists a decimal number e with $\varepsilon > e > 0$, and it is possible to construct a positive integer j_1 such that $m \geq j_1$ and $n \geq j_1$ imply

$$\|x_n - x_m\| < \frac{e}{2} .$$

It is further possible to construct a positive integer j_2 such that $m \geq j_2$ and $n \geq j_2$ imply

$$\|y_n - y_m\| < \frac{e}{2} .$$

We define $j = \max(j_1, j_2)$ and conclude from the triangle inequality for all $m \geq j$ and $n \geq j$

$$\|x_m - y_m\| - \|x_n - y_m\| \leq \|x_m - x_n\| < \frac{e}{2} ,$$

$$\|x_n - y_m\| - \|x_m - y_m\| \leq \|x_n - x_m\| < \frac{e}{2} ,$$

as well as

$$\|x_n - y_m\| - \|x_n - y_n\| \leq \|y_m - y_n\| < \frac{e}{2} ,$$

$$\|x_n - y_n\| - \|x_n - y_m\| \le \|y_n - y_m\| < \frac{e}{2}.$$

Thus

$$|\|x_m - y_m\| - \|x_n - y_m\|| \le \|x_m - x_n\| < \frac{e}{2},$$

$$|\|x_n - y_m\| - \|x_n - y_n\|| \le \|y_m - y_n\| < \frac{e}{2}.$$

From this the triangle inequality of the continuum guarantees for all $m \ge j$ and $n \ge j$

$$|\|x_m - y_m\| - \|x_n - y_n\|| < \varepsilon,$$

and thus proves the assertion by the CAUCHY-criterion. ∎

The most trivial example of a fundamental sequence is the *constant sequence* that only consists of the point x. If X resp. Y denotes the constant sequence consisting only of the point x resp. y, it is clear that $\lim \|X - Y\| = \|x - y\|$. "Par abus de langage" we will denote the constant sequence consisting only of the point x with the same letter x.

Let X, Y, Z denote fundamental sequences. Then we have

$$\lim \|X - Y\| \ge 0$$

and

$$\lim \|Z - X\| - \lim \|Z - Y\| \le \lim \|X - Y\|.$$

Proof. Let X resp. Y resp. Z consist of the points

$$x_1, x_2, \ldots, x_n, \ldots \quad \text{resp.} \quad y_1, y_2, \ldots, y_n, \ldots \quad \text{resp.} \quad z_1, z_2, \ldots, z_n, \ldots.$$

The first formula follows from the fact that, for all positive integers n,

$$\|x_n - y_n\| \ge 0$$

and from the principle of permanence. Further we have for all positive integers n

$$\|z_n - x_n\| - \|z_n - y_n\| \le \|x_n - y_n\|,$$

thus, by the principle of permanence,

$$\lim (\|Z - X\| - \|Z - Y\|) \le \lim \|X - Y\|,$$

which leads, using the fact that

$$\lim (\|Z - X\| - \|Z - Y\|) = \lim \|Z - X\| - \lim \|Z - Y\|,$$

to the second formula. ∎

3.1 Metric spaces and complete metric spaces

General triangle inequality – first version. *Let X denote a fundamental sequence, y, z denote two points, and d', d'' denote two decimal numbers such that*

$$\lim \|X - y\| \leq d' \quad \text{and} \quad \|y - z\| \leq d''.$$

Then we have

$$\lim \|X - z\| \leq d' + d''.$$

Proof. Let X consist of the points $x_1, x_2, \ldots, x_n, \ldots$ and let e be an arbitrary positive decimal number. We can construct a positive integer m such that $n \geq m$ implies that

$$\|x_n - y\| \leq d' + e,$$

and therefore

$$\|x_n - z\| \leq d' + d'' + e.$$

This leads to

$$\lim \|X - z\| \leq d' + d'' + e.$$

As e can be chosen arbitrary small, the assumption $\lim \|X - z\| > d' + d''$ must be absurd. This proves $\lim \|X - z\| \leq d' + d''$. ∎

General triangle inequality – second version. *Let X denote a fundamental sequence, y, z denote two points, and d', d'' denote two decimal numbers such that*

$$\lim \|X - y\| \leq d' \quad \text{and} \quad \lim \|X - z\| \leq d''.$$

Then we have

$$\|y - z\| \leq d' + d''.$$

Proof. Let X consist of the points $x_1, x_2, \ldots, x_n, \ldots$, and let e be an arbitrary positive decimal number. We can construct a positive integer m_1 (resp. a positive integer m_2) such that $n \geq m_1$ (resp. $n \geq m_2$) implies that

$$\|y - x_n\| \leq d' + \frac{e}{2} \quad \text{resp.} \quad \|x_n - z\| \leq d'' + \frac{e}{2}.$$

We set $m = \max(m_1, m_2)$ and thus have for all $n \geq m$

$$\|y - z\| \leq d' + d'' + e.$$

As e can be chosen arbitrary small, the assumption $\|y - z\| > d' + d''$ must be absurd. This proves $\|y - z\| \leq d' + d''$. ∎

General approximation lemma. *Let the fundamental sequence X consist of the points $x_1, x_2, \ldots, x_n, \ldots$, and let $\varepsilon > 0$ denote an arbitrary real number. Then there exists a positive integer j such that, for all $m \geq j$, the inequality*

$$\lim \|X - x_m\| \leq \varepsilon$$

holds.

Proof. It is possible to construct a positive integer j such that, for all $m \geq j$ and all $n \geq j$, the inequality $\|x_n - x_m\| < \varepsilon$ holds. We now let m denote an arbitrary integer with $m \geq j$, and apply the estimation of the limit to the sequence of real numbers consisting of

$$\|x_1 - x_m\|, \|x_2 - x_m\|, \ldots, \|x_n - x_m\|, \ldots.$$

This immediately leads to the assertion. ∎

The propositions above make us look at the system of all fundamental sequences as a metric space in which two fundamental sequences X, Y are defined to be equal if and only if $\lim \|X - Y\| = 0$. We will elaborate upon this later, in a more general setting.

3.1.3 Limit points

Let S denote a metric space. A sequence Σ of sets $S_1, S_2, \ldots, S_n, \ldots$ and a sequence E of positive decimal numbers $e_1, e_2, \ldots, e_2, \ldots$ are called an *adequate pair* (Σ, E) *of sequences* if and only if the following three conditions are fulfilled:
For all positive integers n we have $S_n \subseteq S_{n+1}$ and $e_n > e_{n+1}$.
For each positive integer n, there exists a positive decimal number $d_n < e_n$ such that one can construct, for each point $x \in S$, a point $y \in S_n$ with $\|x - y\| \leq d_n$.
E is convergent with $\lim E = 0$.

Let S denote the system of all decimal numbers, the absolute difference being the metric. Let S_n denote the system of all decimal numbers with exactly n decimal places, and set $e_n = 10^{-n}$. Then, of course, the first and the third of the above conditions are fulfilled. If a denotes an arbitrary decimal number, the decimal number $b = \{a\}_n$ belongs to S_n, and we have

$$|a - b| \leq 5 \times 10^{-n-1} = \frac{e_n}{2}.$$

Thus it is possible to define $d_n = e_n/2$, and the second condition is fulfilled. This example, of course, was our starting point of the construction of the continuum.

Let S denote an arbitrary metric space. Define for any positive integer n the set S_n to coincide with S. Then any sequence E of positive decimal numbers $e_1, e_2, \ldots, e_2, \ldots$ with $e_1 > e_2 > \ldots > e_n > \ldots$ and $\lim E = 0$ together with $\Sigma = (S, S, \ldots, S, \ldots)$ builds an adequate pair (Σ, E): For any positive integer it is possible to choose an arbitrary positive decimal number $d_n < e_n$; for any point $x \in S$ the point $y = x \in S_n = S$ obviously guarantees $\|x - y\| \leq d_n$.
The crucial point of the definition of an adequate pair of sequences, however, is: to gain such a pair with very "small" sets $S_1, S_2, \ldots, S_n, \ldots$. This example therefore is the most trivial and equally most uninteresting special case of this concept.

As a matter of fact, it is unnecessary to restrict the sequence E to positive *decimal* numbers: Suppose we have a sequence E of positive *real* numbers $\varepsilon_1, \varepsilon_2,$

..., ε_2, ... fulfilling the following three conditions:
For all positive integers n we have $\varepsilon_n > \varepsilon_{n+1}$.
For each positive integer n there exists a positive real number $\delta_n < \varepsilon_n$ such that one can construct, for each point $x \in S$, a point $y \in S_n$ with $\|x - y\| \leq \delta_n$.
E is convergent with $\lim E = 0$.
Then we can construct, for any positive integer n, a positive decimal number e'_n such that $\varepsilon_n > e'_n > \varepsilon_{n+1}$, a positive decimal number e''_n such that $\varepsilon_n > e''_n > \delta_n$ and a positive decimal number d_n such that

$$e_n^* = \max\left(e'_n, e''_n\right) > d_n > \delta_n \,.$$

It is clear that the sequence E^* consisting of $e_1^*, e_2^*, \ldots, e_n^*, \ldots$ together with the sequence Σ forms an adequate pair (Σ, E^*). This construction therefore allows us to extend the notion of an adequate pair even to (Σ, E). For the sake of simplicity, we restrict ourselves, however, in the following to the case of sequences E consisting of positive *decimal* numbers.

Let S denote a metric space with (Σ, E) as adequate pair of the two sequences

$$\Sigma = (S_1, S_2, \ldots, S_n, \ldots) \quad \text{and} \quad E = (e_1, e_2, \ldots, e_n, \ldots) \,.$$

A *limit point* ξ is defined to be a sequence consisting of points $[\xi]_1, [\xi]_2, \ldots, [\xi]_n, \ldots$ with the following properties: For any positive integer n the point $[\xi]_n$ belongs to S_n, and for any pair of positive integers n, m the inequality

$$\|[\xi]_n - [\xi]_m\| \leq e_n + e_m$$

holds.

Approximation lemma in metric spaces. *Each limit point ξ is a fundamental sequence, and for any positive integer n the inequality*

$$\lim \|[\xi]_n - \xi\| \leq e_n$$

holds.

Proof. For any real $\varepsilon > 0$ there exists a positive decimal number $e < \varepsilon$ and a positive integer j such that, for all $k \geq j$, the inequality $e_k \leq e/2$ holds. This proves for all integers $n \geq k$ and $m \geq k$

$$\|[\xi]_n - [\xi]_m\| \leq e_n + e_m \leq 2e_k \leq e < \varepsilon \,,$$

i.e. ξ is a fundamental sequence.
We now define, for an arbitrary positive integer n, two sequences A, B as follows:

$$A = \left(\|[\xi]_n - [\xi]_1\|, \|[\xi]_n - [\xi]_2\|, \ldots, \|[\xi]_n - [\xi]_m\|, \ldots\right) ,$$

and

$$B = (e_n + e_1, e_n + e_2, \ldots, e_n + e_m, \ldots) \,.$$

56   3. Metric spaces

The principle of permanence transforms the inequality $A \leq B$ to $\lim A \leq \lim B$. The assertion now follows from $\lim A = \lim \|[\xi]_n - \xi\|$ and $\lim B = e_n$. ∎

Theorem of CAUCHY. *Let* $\Xi = (X_1, X_2, \ldots, X_n, \ldots)$ *denote a sequence of fundamental sequences* $X_1, X_2, \ldots, X_n, \ldots$ *with the following property: For any real* $\varepsilon > 0$ *it is possible to construct a positive integer k such that* $n \geq k$ *and* $m \geq k$ *imply*

$$\lim \|X_n - X_m\| < \varepsilon.$$

Then there exists a limit point ξ *with the following property: For any real* $\varepsilon > 0$ *it is possible to construct a positive integer j such that* $n \geq j$ *implies*

$$\lim \|X_n - \xi\| < \varepsilon.$$

Proof. For any positive integer j it is possible to construct a positive integer k_j such that $n \geq k_j$ and $m \geq k_j$ imply

$$\lim \|X_n - X_m\| < \frac{e_j - d_j}{2}.$$

Let the fundamental sequence X_n consist of the points $x_1^{(n)}, x_2^{(n)}, \ldots, x_m^{(n)}, \ldots$. Since X_{k_j} is a fundamental sequence, we can construct the positive integer l_j with the property

$$\lim \left\| x_{l_j}^{(k_j)} - X_{k_j} \right\| < \frac{e_j - d_j}{2}.$$

Finally it is possible to construct a point $[\xi]_j \in S_j$ such that

$$\left\| [\xi]_j - x_{l_j}^{(k_j)} \right\| \leq d_j.$$

First, we show that the sequence consisting of $[\xi]_1, [\xi]_2, \ldots, [\xi]_n, \ldots$ defines a limit point ξ. For any pair of positive integers n and m it is only necessary to apply the triangle inequality to the five relations

$$\left\| [\xi]_n - x_{l_n}^{(k_n)} \right\| \leq d_n$$

$$\lim \left\| x_{l_n}^{(k_n)} - X_{k_n} \right\| < \frac{e_n - d_n}{2}$$

$$\lim \|X_{k_n} - X_{k_m}\| < \max\left(\frac{e_n - d_n}{2}, \frac{e_m - d_m}{2}\right) < \frac{e_n - d_n}{2} + \frac{e_m - d_m}{2}$$

$$\lim \left\| X_{k_m} - x_{l_m}^{(k_m)} \right\| < \frac{e_m - d_m}{2}$$

$$\left\| x_{l_m}^{(k_m)} - [\xi]_m \right\| \leq d_m.$$

We obtain

$$\|[\xi]_n - [\xi]_m\| \leq d_n + \frac{e_n + d_n}{2}$$
$$+ \left(\frac{e_n - d_n}{2} + \frac{e_m - d_m}{2}\right) + \frac{e_m + d_m}{2} + d_m$$
$$= e_n + e_m .$$

Second, we prove that this limit point ξ fulfills the asserted property of the theorem: Let $\varepsilon > 0$ be an arbitrary real number. We construct a positive decimal number e with $e < \varepsilon$, and construct the positive integer j with the property $e_j \leq e/2$. For any integer $n \geq k_j$ it is possible to combine the four relations

$$\lim \|X_n - X_{k_j}\| < \frac{e_j - d_j}{2}$$
$$\lim \left\|X_{k_j} - x_{l_j}^{(k_j)}\right\| < \frac{e_j - d_j}{2}$$
$$\left\|x_{l_j}^{(k_j)} - [\xi]_j\right\| \leq d_j$$
$$\lim \|[\xi]_j - \xi\| \leq e_j$$

by the triangle inequality to obtain

$$\lim \|X_n - \xi\| \leq \frac{e_j + d_j}{2} + \frac{e_j + d_j}{2} + d_j + e_j = 2e_j \leq e < \varepsilon .$$

This proves the assertion. ∎

3.1.4 Apartness and equality of limit points

Let S denote a metric space with (Σ, E) as adequate pair of the two sequences

$$\Sigma = (S_1, S_2, \ldots, S_n, \ldots) \quad \text{and} \quad E = (e_1, e_2, \ldots, e_n, \ldots) .$$

We define within the system T of all limit points $\xi, \eta, \zeta, \ldots$ an *apartness* relation: $\xi \neq \eta$ holds if and only if $\lim \|\xi - \eta\| > 0$, and an *equality* relation: $\xi = \eta$ holds if and only if $\lim \|\xi - \eta\| = 0$.

The system T of all limit points is a metric space: the distance of two limit points ξ, η is defined to be $\lim \|\xi - \eta\|$. "Par abus de langage" we simply write $\|\xi - \eta\|$ instead of $\lim \|\xi - \eta\|$.

The apartness and the equality within the system T of all limit points ξ, η, ζ fulfill the conditions:
1. *$\xi \neq \eta$ together with $\xi = \eta$ is absurd.*
2. *If $\xi \neq \eta$ is contradictory, then $\xi = \eta$.*
3. *$\xi \neq \xi$ is absurd; we therefore always have $\xi = \xi$.*
4. *$\xi = \eta$ and $\zeta = \eta$ implies that $\xi = \zeta$.*

5. $\xi \neq \eta$ implies, for any point ζ, either that $\xi \neq \zeta$ or that $\eta \neq \zeta$ or both.
6. $\xi = \eta$ implies $\|\xi - \zeta\| = \|\eta - \zeta\|$ and $\|\zeta - \xi\| = \|\zeta - \eta\|$.

Proof. The first three points are obvious by the definition of apartness and equality.
4.: The triangle inequality

$$\|\zeta - \xi\| - \|\zeta - \eta\| \leq \|\xi - \eta\| .$$

together with $\|\xi - \eta\| = 0$ and $\|\zeta - \eta\| = 0$, implies that $\|\xi - \zeta\| = 0$.
5.: Assume $\|\xi - \eta\| > 0$. The interpolation lemma allows us to construct a decimal number d with $\|\xi - \eta\| > d > 0$. The dichotomy lemma guarantees that at least one of the two inequalities

$$\|\zeta - \xi\| < \frac{d}{2} \quad , \quad \|\zeta - \xi\| > \frac{d}{4}$$

holds. Suppose we have

$$\|\zeta - \xi\| < \frac{d}{2} ,$$

and suppose, further, that

$$\|\zeta - \eta\| < \frac{d}{2} ;$$

then the triangle inequality would lead to

$$\|\xi - \eta\| \leq \frac{d}{2} + \frac{d}{2} = d ,$$

which is impossible. Therefore the case $\|\zeta - \xi\| < d/2$ leads to $\zeta \neq \eta$. And the second case $\|\zeta - \xi\| > d/4$ leads to $\zeta \neq \xi$.
6. Assume $\|\xi - \eta\| = 0$. Then the triangle inequality

$$\|\xi - \zeta\| - \|\xi - \eta\| \leq \|\eta - \zeta\|$$

implies that $\|\xi - \zeta\| \leq \|\eta - \zeta\|$. Again the triangle inequality

$$\|\eta - \zeta\| - \|\eta - \xi\| \leq \|\xi - \zeta\|$$

implies that $\|\eta - \zeta\| \leq \|\xi - \zeta\|$. Thus we have the first formula

$$\|\xi - \zeta\| = \|\eta - \zeta\| .$$

The second follows immediately by the symmetry of the metric. ∎

3.1.5 Sequences in metric spaces

An infinite sequence $X = (x_1, x_2, \ldots, x_n, \ldots)$ of points $x_1, x_2, \ldots, x_n, \ldots$ of a metric space S is called *convergent* – more precisely, *convergent in S* – if and only

3.1 Metric spaces and complete metric spaces

if it is possible to construct a point $\xi = \lim X$, the so-called *limit* of the sequence X, with the following property: for any $\varepsilon > 0$ there exists a positive integer j such that, for all $n \geq j$, the inequality $\|\xi - x_n\| < \varepsilon$ holds.

If X is convergent in S, then the relations

$$\xi' = \lim X \quad \text{and} \quad \xi'' = \lim X$$

imply the equality $\xi' = \xi''$.

Proof. Assume $\xi' \neq \xi''$. In other words: $\varepsilon = \|\xi' - \xi''\| > 0$. The interpolation lemma would then allow to construct a decimal number e with $\varepsilon > e > 0$. From $\xi' = \lim X$ we derive the existence of a positive integer j_1 such that, for all $n \geq j_1$, the inequality $\|\xi' - x_n\| < e/2$ holds. From $\xi'' = \lim X$ we derive the existence of a positive integer j_2 such that, for all $n \geq j_2$, the inequality $\|\xi'' - x_n\| < e/2$ holds. This leads for $n \geq \max(j_1, j_2)$ to the contradiction

$$\|\xi' - \xi''\| \leq \frac{e}{2} + \frac{e}{2} = e < \varepsilon = \|\xi' - \xi''\|,$$

i.e. the assumption $\xi' \neq \xi''$ turns out to be absurd. ∎

A sequence $Y = (y_1, y_2, \ldots, y_n, \ldots)$ is called a *subsequence* of the sequence $X = (x_1, x_2, \ldots, x_n, \ldots)$, with notation $Y \sqsubseteq X$, if and only if it is possible to construct a sequence of positive integers $n_1, n_2, \ldots, n_m, \ldots$ such that

$$n_1 < n_2 < \ldots < n_m < \ldots$$

and $y_m = x_{n_m}$ for any positive integer m.

Suppose that the sequence X is convergent in S, then each subsequence Y of X is convergent in S and we have $\lim Y = \lim X$.

Proof. We assume X and Y to be defined as in the definition above. Denote $\lim X = \xi$. To each real $\varepsilon > 0$ there exists a positive integer j such that, for all integers $n \geq j$, the inequality $\|\xi - x_n\| < \varepsilon$ holds. Because $n_m \geq j$ follows from $m \geq j$, we have, a fortiori,

$$\|\xi - y_m\| = \|\xi - x_{n_m}\| < \varepsilon$$

for all $m \geq j$. ∎

Let $X = (x_1, x_2, \ldots, x_n, \ldots)$ and $Y = (y_1, y_2, \ldots, y_n, \ldots)$ denote two sequences. The sequence $Z = X \sqcup Y = (z_1, z_2, \ldots, z_n, \ldots)$ that is constructed by the formulas $z_{2k-1} = x_k$, $z_{2k} = y_k$ for any positive integer k, is the *mixture* of X and Y.

Suppose that the sequences X and Y are convergent in S with $\lim X = \lim Y$, then the mixture $Z = X \sqcup Y$ of X and Y is convergent in S and we have

$$\lim Z = \lim X = \lim Y.$$

Proof. We assume X, Y, and Z to be defined as in the definition above. Denote $\lim X = \lim Y = \zeta$. To each real $\varepsilon > 0$ there exists a positive integer j_1 such that, for all integers $n \geq j_1$, the inequality $\|\zeta - x_n\| < \varepsilon$ holds, and there exists a positive integer j_2 such that, for all integers $n \geq j_2$, the inequality $\|\zeta - y_n\| < \varepsilon$ holds. We now set $j = 2 \max(j_1, j_2)$. Suppose for the integer n that $n \geq j$. In the case that n is odd, $n = 2k - 1$, the inequality $k \geq j_1$ follows from

$$n \geq j \geq 2j_1 \geq 2j_1 - 1,$$

and we have

$$\|\zeta - z_n\| = \|\zeta - x_k\| < \varepsilon.$$

In the case that n is even, $n = 2k$, the inequality $k \geq j_2$ follows from

$$n \geq j \geq 2j_2,$$

and we have

$$\|\zeta - z_n\| = \|\zeta - y_k\| < \varepsilon.$$

∎

3.1.6 Complete metric spaces

A metric space S is called a *complete* metric space if and only if each fundamental sequence is a convergent sequence in S.

Let S denote a metric space with (Σ, E) as adequate pair of the two sequences

$$\Sigma = (S_1, S_2, \ldots, S_n, \ldots) \quad \text{and} \quad E = (e_1, e_2, \ldots, e_n, \ldots),$$

and let T denote the metric space of all limit points of S. Then T is a complete metric space.

Proof. Suppose $(\xi_1, \xi_2, \ldots, \xi_n, \ldots)$ is a fundamental sequence in T. This sequence by definition consists of the fundamental sequences $\xi_1, \xi_2, \ldots, \xi_n, \ldots$ and has the additional property that, for any real $\varepsilon > 0$, it is possible to construct a positive integer k such that $n \geq k$ and $m \geq k$ imply $\|\xi_n - \xi_m\| < \varepsilon$. Then, by CAUCHY's theorem, there exists a limit point ξ, i.e. a point $\xi \in T$, with the following property: For any real $\varepsilon > 0$ it is possible to construct a positive integer j such that $n \geq j$ implies $\|\xi_n - \xi\| < \varepsilon$. ∎

Let S be a metric space. Then there exists a unique complete metric space T that fulfills the following conditions:
1. S is a subspace of T,
2. each point of T is the limit of a fundamental sequence of elements of S.

Proof. Define (Σ, E) to be an arbitrary adequate pair within the metric space S. Then the metric space T consisting of all limit points evidently is a complete metric space with the property that each point of T is limit of a fundamental sequence in S. Of course, S is contained in T. Therefore the existence of T is proved.

Assume that T' is another complete metric space fulfilling the two conditions: first that S is a subspace of T', and second that each point of T' is the limit of a fundamental sequence of elements of S. Then T is a subspace of T', because each point $\xi \in T$ is the limit of a fundamental sequence in S, a fortiori of a fundamental sequence in T', and therefore belonging to T'. But T' itself is a subspace of T, because each limit of a fundamental sequence in S, i.e. each point of T', by CAUCHY's theorem belongs to T. ∎

Let S denote a metric space with (Σ, E) as adequate pair of the two sequences Σ and E, and let T denote the metric space of all limit points of S. Then the set T in the above corollary is determined only by S, and is independent of the choice of the adequate pair (Σ, E). The complete metric space T is called the *completion* of the metric space S.

The completion of a complete metric space coincides with this originally given complete metric space.

3.1.7 Rounded and sufficient approximations

In the following S denotes a metric space with (Σ, E) as adequate pair of the two sequences

$$\Sigma = (S_1, S_2, \ldots, S_n, \ldots) \quad \text{and} \quad E = (e_1, e_2, \ldots, e_n, \ldots),$$

and T denotes the completion of S, i.e. the complete metric space of all limit points.

Rounding lemma. *For each point ξ of the complete metric space T it is possible to construct a point ξ^*, a so-called rounded approximation-sequence of ξ, with the property that the rounded approximations $[\xi^*]_n$ fulfill the inequality*

$$\|[\xi^*]_n - \xi\| \leq \frac{e_n + d_n}{2}$$

for all positive integers n. This of course implies that $\xi^* = \xi$.

Proof. Let n be an arbitrary positive integer. Then there exists a positive integer l_n such that

$$e_{l_n} \leq \frac{e_n - d_n}{2}.$$

Now, $[\xi]_{l_n}$ is a point of S and therefore it is possible to construct a point $[\xi^*]_n$ with the properties $[\xi^*]_n \in S_n$ and

$$\|[\xi^*]_n - [\xi]_{l_n}\| \leq d_n.$$

62   3. Metric spaces

We thus have, for any positive integers n, m, the three inequalities

$$\left\|[\xi^*]_n - [\xi]_{l_n}\right\| \leq d_n,$$
$$\left\|[\xi]_{l_n} - [\xi]_{l_m}\right\| \leq e_{l_n} + e_{l_m} \leq \frac{e_n - d_n}{2} + \frac{e_m - d_m}{2},$$
$$\left\|[\xi]_{l_m} - [\xi^*]_m\right\| \leq d_m,$$

and, by the triangle inequality,

$$\begin{aligned}\left\|[\xi^*]_n - [\xi^*]_m\right\| &\leq d_n + \frac{e_n - d_n}{2} + \frac{e_m - d_m}{2} + d_m \\ &= \frac{e_n + d_n}{2} + \frac{e_m + d_m}{2} \leq e_n + e_m.\end{aligned}$$

This proves that ξ^* belongs to T. Further the two inequalities

$$\left\|[\xi^*]_n - [\xi]_{l_n}\right\| \leq d_n,$$
$$\left\|[\xi]_{l_n} - \xi\right\| \leq e_{l_n} \leq \frac{e_n - d_n}{2}$$

lead to

$$\left\|[\xi^*]_n - \xi\right\| \leq d_n + \frac{e_n - d_n}{2} = \frac{e_n + d_n}{2}$$

as asserted. ∎

Overlapping lemma. *Suppose there exist two points $\xi, \eta \in T$ and a positive integer k such that the inequality*

$$\|\xi - \eta\| \leq \min\left(\frac{e_1 - d_1}{2}, \frac{e_2 - d_2}{2}, \ldots, \frac{e_k - d_k}{2}\right)$$

holds. Define ξ' by setting $[\xi']_n = [\xi^*]_n$ for all integers $n > k$, and by setting at will either $[\xi']_n = [\xi^*]_n$ or $[\xi']_n = [\eta^*]_n$ for positive integers $n \leq k$. Then ξ' is a point of T that coincides with ξ, i.e. $\xi' = \xi$.

Proof. We observe the three inequalities

$$\left\|[\xi^*]_n - \xi\right\| \leq \frac{e_n + d_n}{2}$$
$$\|\xi - \eta\| \leq \min\left(\frac{e_1 - d_1}{2}, \frac{e_2 - d_2}{2}, \ldots, \frac{e_k - d_k}{2}\right)$$
$$\left\|\eta - [\eta^*]_m\right\| \leq \frac{e_m + d_m}{2}$$

the first valid for all positive integers n and the third valid for all positive integers m. In the first case suppose that $n \leq k$ and $m \leq k$. Then $\left\|[\xi']_n - [\xi']_m\right\|$ coincides either with

$$\left\|[\xi^*]_n - [\xi^*]_m\right\| \leq e_n + e_m,$$

or with
$$\|[\eta^*]_n - [\eta^*]_m\| \le e_n + e_m,$$
or with
$$\begin{aligned}\|[\xi^*]_n - [\eta^*]_m\| &\le \frac{e_n + d_n}{2} + \\ &\quad \min\left(\frac{e_1 - d_1}{2}, \frac{e_2 - d_2}{2}, \ldots, \frac{e_k - d_k}{2}\right) + \frac{e_m + d_m}{2} \\ &\le \frac{e_n + d_n}{2} + \left(\frac{e_n - d_n}{2} + \frac{e_m - d_m}{2}\right) + \frac{e_m + d_m}{2} \\ &= e_n + e_m.\end{aligned}$$

In the second case suppose that $n > k$ and $m \le k$. Then $\|[\xi']_n - [\xi']_m\|$ coincides either with
$$\|[\xi^*]_n - [\xi^*]_m\| \le e_n + e_m,$$
or with
$$\begin{aligned}\|[\xi^*]_n - [\eta^*]_m\| &\le \frac{e_n + d_n}{2} + \\ &\quad \min\left(\frac{e_1 - d_1}{2}, \frac{e_2 - d_2}{2}, \ldots, \frac{e_k - d_k}{2}\right) + \frac{e_m + d_m}{2} \\ &\le \frac{e_n + d_n}{2} + \frac{e_m - d_m}{2} + \frac{e_m + d_m}{2} \le e_n + e_m\end{aligned}$$

In the third case suppose that $n > k$ and $m > k$. Then $\|[\xi']_n - [\xi']_m\|$ coincides with
$$\|[\xi^*]_n - [\xi^*]_m\| \le e_n + e_m.$$

Thus we have for all positive integers n and m
$$\|[\xi']_n - [\xi']_m\| \le e_n + e_m,$$

i.e. ξ' is a point of T. The sequence consisting of $[\xi']_1, [\xi']_2, \ldots, [\xi']_n, \ldots$ has the same limit as the sequence consisting of $[\xi^*]_1, [\xi^*]_2, \ldots, [\xi^*]_n, \ldots$. We therefore deduce $\xi' = \xi^*$ and from that $\xi' = \xi$. ∎

For example, let $\xi \in T$ denote an arbitrary point and k an arbitrary positive integer. Then there exists a positive integer n such that
$$e_n \le \min\left(\frac{e_1 - d_1}{2}, \frac{e_2 - d_2}{2}, \ldots, \frac{e_k - d_k}{2}\right).$$

The approximation $[\xi]_n$ of ξ can be interpreted as a point $x = [\xi]_n$ of T, and we may assume that $[x]_1, [x]_2, \ldots, [x]_k$ are the rounded approximations of $[\xi]_n$. We call $x = [\xi]_n$ a *sufficient approximation of* ξ (with respect to $[\xi]_1, [\xi]_2, \ldots,$

$[\xi]_k$, more precisely therefore: a *k-sufficient approximation of* ξ), because the inequality
$$\|x - \xi\| = \|[\xi]_n - \xi\| \leq e_n$$
and the overlapping lemma guarantee the following fact: If ξ' is defined by
$$[\xi']_j = \begin{cases} [x]_j & \text{for all } j \leq k \\ [\xi^*]_j & \text{for all } j > k, \end{cases}$$
then ξ' is a point of T that coincides with ξ, i.e. $\xi' = \xi$. In other words: It is possible to exchange the approximations $[\xi]_1, [\xi]_2, \ldots, [\xi]_k$ by the rounded approximations $[x]_1, [x]_2, \ldots, [x]_k$ of the sufficient approximation $x = [\xi]_n$, without changing the point ξ itself.

3.2 Compact metric spaces

3.2.1 Bounded and totally bounded sequences

A sequence X of a metric space S is called *bounded* if and only if it is possible to construct a point z and a decimal number a such that, for all points $x \in X$, the inequality $\|x - z\| \leq a$ holds.

A sequence X of a metric space S is bounded if and only if it is possible to construct to any point y a decimal number c such that, for all points $x \in X$, the inequality $\|x - y\| \leq c$ holds.

Proof. It is clear that the condition of the proposition guarantees the boundness of X. Assume on the other hand that X is a bounded sequence, i.e. that the point z and the decimal number a are constructed such that, for all points $x \in X$, the inequality $\|x - z\| \leq a$ holds. Let y denote an arbitrary point. There exists a decimal number b with the property $\|y - z\| \leq b$. We define $c = a + b$ and conclude from the two inequalities $\|x - z\| \leq a$ and $\|y - z\| \leq b$ the inequality $\|x - y\| \leq c$ for all points $x \in X$. ∎

A finite sequence $(x_1, x_2, \ldots, x_n)$ of points $x_1, x_2, \ldots, x_n$ that belong to a sequence X of a metric space S is called a *finite ε-net* of X if and only if it is possible to construct to each point $x \in X$ a positive integer $j \leq n$ such that $\|x - x_j\| < \varepsilon$.

A sequence X of a metric space S is called *totally bounded* if and only if there exists, for any real $\varepsilon > 0$, a finite ε-net of X.

Each totally bounded sequence is a bounded sequence.

Proof. Let X denote a totally bounded sequence and let $(x_1, x_2, \ldots, x_n)$ denote a finite 1-net of X. For each integer j with $1 \leq j \leq n$ we choose a decimal number a_j such that $\|x_j - x_1\| \leq a_j$. Finally we set $z = x_1$ and
$$a = \max(a_1 + 1, a_2 + 1, \ldots, a_n + 1).$$

For any point $x \in X$ it is possible to construct a positive integer $j \leq n$ such that $\|x - x_j\| < 1$, a fact that together with $\|x_j - x_1\| \leq a_j$ leads to

$$\|x - z\| = \|x - x_1\| \leq a_j + 1 \leq a.$$

∎

If X and Y are totally bounded sequences, then $X \sqcup Y$ is also totally bounded.

Proof. Let $\varepsilon > 0$ denote an arbitrary real number. X resp. Y allow to construct a finite ε-net $(x_1, x_2, \ldots, x_n)$ resp. $(y_1, y_2, \ldots, y_m)$. The finite sequence

$$(z_1, z_2, \ldots, z_{n+m}) = (x_1, x_2, \ldots, x_n) \sqcup (y_1, y_2, \ldots, y_m)$$

is defined by setting

$$(z_1, z_2, \ldots, z_{n+m}) = \begin{cases} (x_1, y_1, \ldots, x_k, y_k, y_{k+1}, \ldots, y_m) \\ \quad \text{if } k = \min(n, m) = n, \\ (x_1, y_1, \ldots, x_k, y_k, x_{k+1}, \ldots, x_n) \\ \quad \text{if } k = \min(n, m) = m. \end{cases}$$

This finite sequence forms a finite ε-net of $X \sqcup Y$. The reason is that, for any point $z \in X \sqcup Y$, at least one of the equalities $z = x$ or $z = y$ with points $x \in X$ or $y \in Y$ respectively must hold. In the first case, there exists a positive integer $j \leq n$ such that $\|x - x_j\| < \varepsilon$, and in the second case, there exists a positive integer $l \leq m$ such that $\|y - y_l\| < \varepsilon$. Both x_j and y_l belong to the sequence $(z_1, z_2, \ldots, z_{n+m})$. ∎

A sequence of points that is convergent in the metric space is totally bounded.

Proof. Let the sequence X consist of the points $x_1, x_2, \ldots, x_n, \ldots$ and assume that X converges in the metric space S with limit $\xi = \lim X$. Let $\varepsilon > 0$ be an arbitrary real number and construct a decimal number e such that $\varepsilon > e > 0$. There exists a positive integer n such that, for all $m \geq n$, the inequality $\|x_m - \xi\| \leq e/2$ holds; in particular, $\|x_n - \xi\| \leq e/2$. We therefore have for all integers $m \geq n$ the relation $\|x_m - x_n\| \leq e$, which proves that $(x_1, x_2, \ldots, x_n)$ forms a finite ε-net of X. ∎

3.2.2 Located sequences

A sequence X of a metric space S is called *located* if and only if it is possible, for any given point ξ and any given real numbers α, β with $\alpha > \beta$,
either to construct a point $x_0 \in X$ such that $\|x_0 - \xi\| < \alpha$,
or to prove the inequality $\|x - \xi\| > \beta$ for all points $x \in X$.

The possibility that both assertions are valid is of course not excluded. It is further enough to restrict the definition to *decimal numbers* a, b with $a > b$ instead of real numbers α, β, i.e.

*A sequence X of a metric space S is called located if and only if it is possible, for any given point ξ and any given decimal numbers a, b with $a > b$,
either to construct a point $x_0 \in X$ such that $\|x_0 - \xi\| < a$,
or to prove the inequality $\|x - \xi\| > b$ for all points $x \in X$.*

Proof. A located sequence X of course has this property. Now, we prove the converse. Assume that X has the announced property. Let α, β be arbitrary real numbers with $\alpha > \beta$. Then, by the interpolation lemma, there exist two decimal numbers a, b such that

$$\alpha > a > b > \beta.$$

On the one hand, the inequality $\|x_0 - \xi\| < \alpha$ follows from $\|x_0 - \xi\| < a$. On the other hand, a proof that $\|x - \xi\| > b$ is valid for all points $x \in X$ immediately proves that $\|x - \xi\| > \beta$ is valid for all points $x \in X$. ∎

*Let T denote the completion of the metric space S. A sequence X of T is located if and only if it is possible, for any given point $y \in S$ and any given decimal numbers a, b with $a > b$,
either to construct a point $x_0 \in X$ such that $\|x_0 - y\| < a$,
or to prove the inequality $\|x - y\| > b$ for all points $x \in X$.*

Proof. It is clear that a located sequence fulfills the condition of the proposition. Assume on the other hand that (Σ, E) with E consisting of $e_1, e_2, \ldots, e_n, \ldots$ is an adequate pair, and let ξ denote an arbitrary point of T and a, b decimal numbers with $a > b$. We then define the integer n so big that

$$e_n < \frac{a-b}{4},$$

the decimal numbers a', b' by the formulas

$$a' = \frac{3a+b}{4}, \quad b' = \frac{a+3b}{4}$$

that guarantee

$$a > a' + e_n > a' > b' > b' - e_n > b,$$

and the point $y \in S$ by setting $y = [\xi]_n$. Suppose it is possible to construct a point $x_0 \in X$ such that $\|x_0 - y\| < a'$. Then this, together with

$$\|y - \xi\| = \|[\xi]_n - \xi\| \leq e_n,$$

leads to

$$\|x_0 - \xi\| \leq a' + e_n < a.$$

Suppose there exists a proof that $\|x - y\| > b'$ for all $x \in X$. Then again this, together with

$$\|y - \xi\| = \|[\xi]_n - \xi\| \leq e_n,$$

leads to

$$\|x - \xi\| \geq b' - e_n > b,$$

as asserted. ∎

Each totally bounded sequence is located.

Proof. Let X be totally bounded. ξ designates an arbitrary point and a, b designate two arbitrary decimal numbers with $a > b$. We define $e = (a-b)/4$, and we are able to construct a finite e-net $(x_1, x_2, \ldots, x_n)$ of X. The dichotomy lemma allows, for each positive integer $j \leq n$, to accept at least one of the following inequalities

$$\|x_j - \xi\| < a \quad \text{or} \quad \|x_j - \xi\| > a - e.$$

Therefore at least one of the following two cases must take place: In the first case there exists a positive integer $j \leq n$ such that $\|x_j - \xi\| < a$. In the second case we have $\|x_j - \xi\| > a - e$ for all positive integers $j \leq n$. If this second case occurs, the inequality $\|x - \xi\| < a - 2e$ is absurd for all $x \in X$, because otherwise it would be possible to find a point x_j from the e-net of X such that $\|x_j - x\| < e$, which together with $\|x - \xi\| < a - 2e$ would lead to

$$\|x_j - \xi\| < (a - 2e) + e = a - e$$

– the inequality that the second case excludes. In this second case, we therefore have

$$\|x - \xi\| \geq a - 2e > b$$

for all $x \in X$. ∎

3.2.3 The infimum

Let S be a metric space, X be a sequence of points and ξ be a point. A real number $\mu = \inf \|X - \xi\|$ is called the *infimum of* $\|X - \xi\|$ if and only if this real number fulfills the following two conditions:
μ is a *lower bound* of $\|X - \xi\|$, i.e. we have $\mu \leq \|x - \xi\|$ for all $x \in X$,
μ is the *greatest* lower bound of $\|X - \xi\|$, i.e. for all real numbers $\lambda > \mu$ it is possible to find a point $x_0 \in X$ such that $\lambda > \|x_0 - \xi\|$.

Suppose that the infimum of $\|X - \xi\|$ exists. Then this real number is uniquely defined.

Proof. The inequality $\lambda > \mu$ obviously forbids that both real numbers λ and μ be infima of $\|X - \xi\|$. ∎

Theorem about the infimum. *The infimum of $\|X - \xi\|$ exists for all points ξ if and only if X is a located sequence.*

Proof. Suppose first that ξ designates an arbitrary point and the real number $\mu = \inf \|X - \xi\|$ exists. Let α, β be arbitrary real numbers with $\alpha > \beta$. The dichotomy lemma guarantees that at least one of the inequalities $\alpha > \mu$ or $\mu > \beta$ is valid. In the case $\alpha > \mu$, by definition of μ, it is possible to construct a point

$x_0 \in X$ such that $\alpha > \|x_0 - \mu\|$. In the case $\mu > \beta$, again by definition of μ, we have

$$\beta < \mu \leq \|x - \xi\|$$

for all $x \in X$. Therefore X is seen to be located.

Now suppose that X is a located sequence and let ξ denote an arbitrary point. Let x_1 be the first point of X and define w_0 to be a positive integer with the property $w_0 > \|x_1 - \xi\|$. As X is located, for each integer w at least one of the two cases
case 1: there exists a $x_0 \in X$ such that $w > \|x_0 - \xi\|$,
case 2: for all $x \in X$ the inequality $\|x - \xi\| > w - 1$ holds,
must take place.

In particular case 1 takes place if $w = w_0$, because we may choose $x_0 = x_1$. And case 2 takes place if $w = 0$, because the metric is positive. Therefore it is possible to establish the following procedure:

The procedure starts with $w = 0$. At this starting point, case 1 is absurd.

Suppose the procedure has already reached the integer $w \geq 0$. If case 2 is realized for this w, then w is exchanged by $w + 1$. If for $w + 1$ instead of w case 1 is realized, then the procedure stops. It is clear that the procedure will stop anyhow – at the latest if it has reached $w = w_0 - 1$.

The procedure therefore announces a nonnegative integer z such that on one hand we have $\|x - \xi\| > z - 1$ for all $x \in X$, and on the other hand there exists a $x_0 \in X$ with the property $\|x_0 - \xi\| < z + 1$.

We now assume that, for some positive integer n, we have already constructed a decimal number

$$[\mu]_{n-1} = z + w_1 \times 10^{-1} + \ldots + w_{n-1} \times 10^{-n+1}$$

with integers $w_1, \ldots, w_{n-1}$ such that, for all positive integers $j < n$, we have $-9 \leq w_j \leq 9$, with the following property: on one hand we have

$$\|x - \xi\| > [\mu]_{n-1} - 10^{-n+1}$$

for all $x \in X$, and on the other hand there exists a $x_0 \in X$ with the property

$$\|x_0 - \xi\| < [\mu]_{n-1} + 10^{-n+1} .$$

We then repeat an analogous procedure with the decimal number

$$z + w_1 \times 10^{-1} + \ldots + w_{n-1} \times 10^{-n+1} + w \times 10^{-n} :$$

As X is located, for each integer w at least one of the two cases
case 1: there exists a $x_0 \in X$ such that

$$z + w_1 \times 10^{-1} + \ldots + w_{n-1} \times 10^{-n+1} + w \times 10^{-n} > \|x_0 - \xi\| ,$$

case 2: for all $x \in X$ we have the inequality

$$\|x - \xi\| > z + w_1 \times 10^{-1} + \ldots + w_{n-1} \times 10^{-n+1} + (w - 1) \times 10^{-n} ,$$

must take place.
In particular case 1 takes place if $w = 10$, and case 2 takes place if $w = -9$, because this was the induction hypothesis. Therefore the procedure starts with $w = -9$. At this starting point, case 1 is absurd.
Suppose the procedure has already reached the integer $w \geq -9$. If case 2 is realized for this w, then w is exchanged by $w + 1$. If for $w + 1$ instead of w case 1 is realized, then the procedure stops. It is clear that the procedure will stop anyhow – at the latest if $w = 9$.
The procedure therefore announces an integer w_n that guarantees the following: The decimal number

$$[\mu]_n = z + w_1 \times 10^{-1} + \ldots + w_{n-1} \times 10^{-n+1} + w_n \times 10^{-n}$$

(with integers $w_1, \ldots, w_{n-1}, w_n$ such that, for all positive integers $j \leq n$, the relation $-9 \leq w_j \leq 9$ holds) has the following property: on one hand we have

$$\|x - \xi\| > [\mu]_n - 10^{-n}$$

for all $x \in X$, and on the other hand there exists a $x_0 \in X$ with the property

$$\|x_0 - \xi\| < [\mu]_n + 10^{-n} .$$

The sequence μ consisting of $[\mu]_1, [\mu]_2, \ldots, [\mu]_n, \ldots$ defines a pendulum number, i.e. a real number μ. We now prove that μ fulfills the conditions of the infimum of $\|X - \xi\|$:
On the one hand, we have for all points $x \in X$ and all positive integers n

$$\|x - \xi\| > [\mu]_n - 10^{-n} , \quad \text{i.e.} \quad [\mu]_n - \|x - \xi\| < 10^{-n} ,$$

which, by the principle of permanence, implies that $\mu - \|x - \xi\| \leq 0$. Thus for all $x \in X$ we have

$$\mu \leq \|x - \xi\|$$

i.e. μ is a lower bound of $\|X - \xi\|$.
On the other hand, assume $\lambda > \mu$. By the interpolation lemma there exists a positive integer n such that $2 \times 10^{-n} \leq \lambda - \mu$. As the approximation lemma proves

$$\left([\mu]_n + 10^{-n}\right) - \mu \leq \left|\left([\mu]_n + 10^{-n}\right) - \mu\right| \leq 2 \times 10^{-n} \leq \lambda - \mu$$

we even have

$$[\mu]_n + 10^{-n} \leq \lambda .$$

Now we can find a point $x_0 \in X$ fulfilling

$$[\mu]_n + 10^{-n} > \|x_0 - \xi\| ,$$

and by this we have produced an example of a point $x_0 \in X$ with $\lambda > \|x_0 - \xi\|$.
∎

3.2.4 The hypothesis of DEDEKIND and CANTOR

The *hypothesis of* DEDEKIND *and* CANTOR postulates: all sequences of all metric spaces are located.

Assume that the metric space S only consists of the two numbers 0 and 1 with the absolute difference as metric. Let X denote the sequence of the numbers x_1, $x_2, \ldots, x_n, \ldots$ within this space, i.e. each of the numbers $x_1, x_2, \ldots, x_n, \ldots$ is either zero or one. If X is located, then we can decide between the following two possibilities:
either all numbers in X are zero,
or there is at least one number in X that coincides with one.
(The proof of this implication follows immediately from the definition of locatedness by setting $\xi = 1, \alpha = 3/4, \beta = 1/4$.)

Although the implication of this example at first sight seems to be plausible, this implication contradicts the intuitive concept of infinity: It is obvious that the implication is true if X were a *finite* sequence, as we are, in principle, able to "run through" all members $x_1, x_2, \ldots, x_n$ of a sequence consisting of n elements, and to decide for each of these if it coincides with zero or one. But by the very meaning of "infinity", this imagination of "running through" is impossible if X is an infinite sequence.

As a matter of fact, the notion of "sequence" given so far does *not* fix the appropriate use of this concept. To do this, it is necessary to put this notion into a precise form. We here, following an idea of LORENZEN, establish that frame in form of a dialogue of two intellectual combatants, of the so-called *proponent* and the so-called *opponent*. These two partners first have to reach an agreement as to the *logic* they are going to use – the proponent in putting forward and defending propositions and the opponent in casting doubt on all the proponent's assertions that are not yet proved.

If both the proponent and the opponent accept the hypothesis of DEDEKIND and CANTOR, they thereby agree that, by setting the sequence X, *all* its elements x_1, $x_2, \ldots, x_n, \ldots$ are given *at once*. Beyond that, the *whole* infinite sequence $x_1, x_2, \ldots, x_n, \ldots$ is at disposal. Only this precondition – BISHOP called it very appropriately a "principle of omniscience" – permits the proponent to defend the assertion that within the metric space consisting only of 0 and 1 either all numbers in X are zero, or there is at least one number in X that coincides with one. He banishes the doubts of the opponent by demanding his partner to set an arbitrary sequence X. Along with X the *whole* sequence of integers $x_1, x_2, \ldots, x_n, \ldots$ is announced by the opponent, which implies that the decision between the two alternatives is as trivial as it is in the case of a finite sequence X.

It is evident that if we take the notion of infinity seriously, we are forced to reject "principles of omniscience" and arguments like these, and we acknowledge this for two different reasons:

In *recursive constructivism* – a branch of mathematics that we only mention here, but do not follow up further – the indication of a sequence X proceeds with the indication of the *recursive function* that produces for any n the n-th element

x_n of this sequence. The assertion of the proponent that within the metric space consisting only of 0 and 1 either all numbers in X are zero, or there is at least one number in X that coincides with one, should be understood as if the proponent claimed to be able to distinguish between these possibilities by means of recursive methods – a totally untenable assertion.

BROUWER's *intuitionism* goes even further: The indication of a sequence X by one partner of the dialogue proceeds with the mere indication of the n-th element x_n of X if the other partner of the dialogue wants to know this n-th element. The "method" – whatever this vague notion might mean – by which the presenting partner obtains this point x_n *is no topic of the dialogue at all*. The assertion of the proponent that within the metric space consisting only of 0 and 1 either all numbers in X are zero, or there is at least one number in X that coincides with one, is rejected by the opponent by setting a sequence X: for each natural number n that the proponent submits, the opponent announces $x_n = 0$ as the n-th element of X. But this certainly does not imply that *all* numbers $x_1, x_2, \ldots, x_n, \ldots$ of X are zero. Thus the proponent cannot win this dispute.

Although the acceptance of the hypothesis of DEDEKIND and CANTOR would substantially simplify many of the proofs given up to this stage, as well as many of the proofs in the following text, we decide to *reject* this hypothesis and to adopt the position of BROUWER's *intuitionism*: all proofs have to be read as records of dialogues between the proponent and the opponent under the assumption that the indication of a sequence by one partner of the dialogue proceeds with the *mere* indication of its n-th element if the other partner of the dialogue wants to know this n-th element for any positive integer n.

3.2.5 Bounded, totally bounded, and located sets

A set U of a metric space S is called *bounded* if and only if it is possible to construct a point z and a decimal number a such that, for all points $x \in U$, the inequality $\|x - z\| \leq a$ holds.

A finite sequence $(x_1, x_2, \ldots, x_n)$ of points $x_1, x_2, \ldots, x_n$ that belong to a set U of a metric space S is called a *finite ε-net* of U if and only if it is possible to construct to each point $x \in U$ a positive integer $j \leq n$ such that $\|x - x_j\| < \varepsilon$.

A set U of a metric space S is called *totally bounded* if and only if there exists, for any real $\varepsilon > 0$, a finite ε-net of U.

A set U of a metric space S is called *located* if and only if it is possible, for any given point ξ and any given real numbers α, β with $\alpha > \beta$,
either to construct a point $x_0 \in U$ such that $\|x_0 - \xi\| < \alpha$,
or to prove the inequality $\|x - \xi\| > \beta$ for all points $x \in U$.

Let S be a metric space, U be a set and ξ be a point. A real number

$$\mu = \inf \|U - \xi\|$$

is called the *infimum of* $\|U - \xi\|$ if and only if this real number fulfills the following two conditions:

μ is a *lower bound of* $\|U - \xi\|$, i.e. we have $\mu \leq \|x - \xi\|$ for all $x \in U$,
μ is the *greatest* lower bound of $\|U - \xi\|$, i.e. for all real numbers $\lambda > \mu$ it is possible to find a point $x_0 \in U$ such that $\lambda > \|x_0 - \xi\|$.

All five definitions above differ from the analogous definitions of the same concepts by exchanging the term "sequence" by "set" and the accompanying letter "X" by "U". The same exchange transfers the proofs of the foregoing assertions to proofs of the following corollaries:

A set U of a metric space S is bounded if and only if it is possible to construct to any point y a decimal number c such that, for all points $x \in U$, the inequality $\|x - y\| \leq c$ holds.

Each totally bounded set is a bounded set.

If U and V are totally bounded sets, their union $U \cup V$ is also totally bounded.

*Let T denote the completion of the metric space S. A set U of T is located if and only if it is possible, for any given point $y \in S$ and any given decimal numbers a, b with $a > b$,
either to construct a point $x_0 \in U$ such that $\|x_0 - y\| < a$,
or to prove the inequality $\|x - y\| > b$ for all points $x \in U$.*

We consider, as an example, the set of all decimal numbers as metric space with the absolute difference as its metric. Let j denote a positive integer. The set of all decimal numbers with exactly j decimal places is a located set in the continuum.

Each totally bounded set is located.

Suppose that the infimum of $\|U - \xi\|$ exists. Then this real number is uniquely defined.

Set-theoretic version of the theorem about the infimum. *The infimum of $\|U - \xi\|$ exists for all points ξ if and only if U is a located set.*

3.2.6 Separable and compact spaces

Let S denote a metric space with (Σ, E) as adequate pair of the two sequences

$$\Sigma = (S_1, S_2, \ldots, S_n, \ldots) \quad \text{and} \quad E = (e_1, e_2, \ldots, e_n, \ldots),$$

and let T denote the metric space of all limit points of S. If S is given as a sequence, then T is called a *separable* metric space. If, furthermore, for each positive integer n the subset S_n of S is given as a finite sequence, then T is called a *compact* metric space.

A complete metric space is a compact metric space if and only if it is a totally bounded metric space.

Proof. First assume that T denotes a compact metric space, and we adopt the notation from the definition above. Let $\varepsilon > 0$ denote an arbitrary real number.

Define the positive integer n so big that $e_n < \varepsilon$. There exists a positive decimal number $d_n < e_n$ such that one can construct to each point $x \in S$ a point $y \in S_n$ with
$$\|x - y\| \le d_n < e_n < \varepsilon .$$
This proves that S_n forms a finite ε-net of S.

Furthermore we can construct a positive integer m such that $e_m < e_n - d_n$. If ξ denotes an arbitrary point of T, we define $x = [\xi]_m$, a point belonging to S, construct as before the point $y \in S_n$, and conclude from
$$\|\xi - x\| = \big\|\xi - [\xi]_m\big\| \le e_m < e_n - d_n$$
together with $\|x - y\| \le d_n$ that
$$\|\xi - y\| \le e_n < \varepsilon ,$$
i.e. S_n is a finite ε-net of T.

Second assume that T denotes a totally bounded complete metric space, and let D denote a sequence of positive decimal numbers $d_1, d_2, \ldots, d_n, \ldots$ with
$$d_1 > d_2 > \ldots > d_n > \ldots \quad \text{and} \quad \lim D = 0 ,$$
and define for each positive integer n the decimal number $e_n = 2d_n$. The sequence E consisting of $e_1, e_2, \ldots, e_n, \ldots$ also has the property
$$e_1 > e_2 > \ldots > e_n > \ldots \quad \text{and} \quad \lim E = 0 .$$

The prerequisite allows us to construct, for each positive integer n, a finite sequence X_n of points for which it is possible to find to each point $\xi \in T$ a point $x \in X_n$ such that $\|x - \xi\| < d_n$. The set S_n consisting of the points belonging to $X_1 \sqcup X_2 \sqcup \ldots \sqcup X_n$ also is finite, and we have the chain of inclusions
$$S_1 \subseteq S_2 \subseteq \ldots \subseteq S_n \subseteq \ldots .$$

The metric space S consisting of all points that appear in at least one of the X_n has of course the property that, to each point $x \in S$ and to each positive integer n, there exists a point $y \in S_n$ fulfilling $\|x - y\| \le d_n$. Thus the sequence Σ consisting of $S_1, S_2, \ldots, S_n, \ldots$ together with E forms an adequate pair (Σ, E), and it is easy to prove that T is the completion of S: On the one hand, S is a subspace of T, and on the other hand, for each point ξ of T and for each positive integer n it is possible to find a point $x_n \in S_n$ such that $\|\xi - x_n\| \le d_n$; this proves that the sequence of the points $x_1, x_2, \ldots, x_n, \ldots$ belonging to S converges to ξ, i.e. each point of T is limit of a fundamental sequence in S. ∎

A set in a compact metric space is located if and only if it is totally bounded.

Proof. It is already clear that a totally bounded set is located. Suppose on the other hand that U denotes a located set in the compact space T and let $\varepsilon > 0$ denote an arbitrary real number. We construct a decimal number e with $\varepsilon > e > 0$, and we construct further a finite $e/4$-net $(\xi_1, \xi_2, \ldots, \xi_n)$ of T. For each positive integer $j \leq n$ we can state that at least one of the inequalities

$$\inf \|U - \xi_j\| > \frac{e}{4} \quad \text{or} \quad \inf \|U - \xi_j\| < \frac{e}{2}$$

holds. We therefore can divide the set of all positive integers $j \leq n$ into two disjoint subsets J' and J'' such that $j \in J'$ implies that $\inf \|U - \xi_j\| > e/4$ and $j \in J''$ implies that $\inf \|U - \xi_j\| < e/2$. It is clear that, for each $x \in U$, there exists a positive integer $j \leq n$ such that $\|x - \xi_j\| < e/4$. This inequality of course excludes the possibility $j \in J'$, the relation $j \in J''$ therefore is a necessary consequence. By definition of the infimum, to each $j \in J''$ there must exist a $x_j \in U$ such that $\|x_j - \xi_j\| < 3e/4$. This implies by the triangle inequality that

$$\|x - x_j\| \leq \frac{e}{4} + \frac{3e}{4} = e < \varepsilon$$

and thus proves that the finite sequence consisting of all x_j with $j \in J''$ forms a finite ε-net of U. ∎

3.2.7 Bars

Let S denote a metric space with (Σ, E) as adequate pair of the two sequences

$$\Sigma = (S_1, S_2, \ldots, S_n, \ldots) \quad \text{and} \quad E = (e_1, e_2, \ldots, e_n, \ldots) \,,$$

and let T denote the metric space of all limit points of S. A subset W of S is called a *bar* if and only if it is possible to detect for each point $\xi \in T$ a positive integer k such that, for each integer $n \geq k$, the set W contains the point $[\xi]_n$.

Let, as a first example, S_n denote the space of all decimal numbers a of the form

$$a = 0.z_1 z_2 \ldots z_n = z_1 \times 10^{-1} + z_2 \times 10^{-2} + \ldots + z_n \times 10^{-n} \,,$$

$z_1, z_2, \ldots, z_n$ being integers with $0 \leq z_j \leq 9$ for all positive integers $j \leq n$. Define S as the union of $S_1, S_2, \ldots, S_n, \ldots$, i.e. the set of all decimal numbers a fulfilling $0 \leq a < 1$. The metric on S is the absolute difference and the sequence E consists of the numbers $e_n = 10^{-n}$ for all positive integers n. The completion T of S consists of all real numbers α fulfilling $0 \leq \alpha \leq 1$.

Let m denote a positive integer, and let W be the set of all decimal numbers

$$a = 0.z_1 z_2 \ldots z_n$$

for which either at least one of the digits $z_1, z_2, \ldots, z_n$ is different from zero, or $n \geq m$ holds. W obviously is a bar.

The following *procedure* detects for any $\alpha \in T$ a positive integer k such that, for each integer $n \geq k$, the bar W contains the point $[\alpha]_n$: We consider that the decimal numbers $[\alpha]_1$, $[\alpha]_2$, ..., $[\alpha]_j$ are submitted, and observe at the moment the case $j < m$. If we find a positive integer $k \leq j$ such that $[\alpha]_k > 10^{-k}$, then we can be sure that the possibility $[\alpha]_n = 0$ is ruled out for all $n \geq k$, i.e. the bar W contains the point $[\alpha]_n$ for all $n \geq k$. If otherwise we have $[\alpha]_n \leq 10^{-n}$ for all positive integers $n \leq j$, the procedure requires the datum of the next decimal number $[\alpha]_{j+1}$ in order, either to register that $[\alpha]_{j+1} > 10^{-j-1}$, i.e. that for each integer $n \geq j + 1$ the bar W contains the point $[\alpha]_n$, or again to have $[\alpha]_{j+1} \leq 10^{-j-1}$ and to require the datum of the next decimal number, or, finally, to register that the level $j + 1 = k = m$ is attained, in which case it is clear by definition of W that for each integer $n \geq k$ the bar W contains the point $[\alpha]_n$.

Let, as a second example, S_n denote the space of all decimal numbers a of the form

$$a = z + 0.z_1 z_2 \ldots z_n = z + z_1 \times 10^{-1} + z_2 \times 10^{-2} + \ldots + z_n \times 10^{-n},$$

z being an integer and z_1, z_2, ..., z_n being integers with $0 \leq z_j \leq 9$ for all positive integers $j \leq n$. Define S as the union of S_1, S_2, ..., S_n, ..., i.e. the set of all decimal numbers. The metric on S is the absolute difference and the sequence E consists of the numbers $e_n = 10^{-n}$ for all positive integers n. The completion T of S consists of all real numbers.

Let m denote a positive integer, and let W be the set of all decimal numbers $a = z + 0.z_1 z_2 \ldots z_n$ for which either at least one of the digits z_1, z_2, ..., z_n is different from zero, or $n \geq m + |z|$ holds. W obviously is a bar.

The following *procedure* detects for any $\alpha \in T$ a positive integer k such that, for each integer $n \geq k$, the bar W contains the point $[\alpha]_n$: We consider that the decimal numbers $[\alpha]_1 = z + 0.z_1$, $[\alpha]_2$, ..., $[\alpha]_j$ are submitted, and observe at the moment the case $j < m + |z|$. If we find a positive integer $k \leq j$ such that $z + 10^{-k} < [\alpha]_k < z + 1 - 10^{-k}$, then we can be sure that the possibilities $[\alpha]_n = z$ and $[\alpha]_n = z + 1$ are ruled out for all $n \geq k$, i.e. the bar W contains the point $[\alpha]_n$ for all $n \geq k$. If otherwise we have $z_1 \leq 1$ and $[\alpha]_n \leq z + 10^{-n}$ for all positive integers $n \leq j$, or $z_1 = 9$ and $[\alpha]_n \geq z + 1 - 10^{-n}$ for all positive integers $n \leq j$, the procedure requires the datum of the next decimal number $[\alpha]_{j+1}$ in order, either to register that $z + 10^{-j-1} < [\alpha]_{j+1} < z + 1 - 10^{-j-1}$, i.e. that for each integer $n \geq j + 1$ the bar W contains the point $[\alpha]_n$, or otherwise again to require the datum of the next decimal number, or, finally, to register that the level $j + 1 = k = m + |z|$ or $j + 1 = k = m + |z + 1|$ according to $z_1 \leq 1$ or $z_1 = 9$ respectively is attained, in which case it is clear by definition of W that for each integer $n \geq k$ the bar W contains the point $[\alpha]_n$.

These two examples are paradigmatic for the abstract concept of a bar. Although they seem to be rather similar, they essentially differ by the fact that in the first example a level m from which on in any case the bar W contains the point $[\alpha]_n$ for all $n \geq m$ can be constructed, whereas in the second example such an "universal level" does not exist. This is the main issue of the following theorem.

3.2.8 Bars and compact spaces

Bar-theorem. *Suppose that T is a compact metric space. Let S denote a metric space with (Σ, E) as adequate pair of the two sequences*

$$\Sigma = (S_1, S_2, \ldots, S_n, \ldots) \quad \text{and} \quad E = (e_1, e_2, \ldots, e_n, \ldots),$$

such that T is the completion of S, and all S_1, S_2, ..., S_n, ... are finite. Let the subset W of S denote a bar. Then it is possible to detect a positive integer m such that, for all points $\xi \in T$ and all positive integers $n \geq m$, the points $[\xi]_n$ belong to W.

Proof. Since we *know* that W is a bar, we must have a *procedure* that detects for any point $\xi \in T$ a positive integer k such that, for each integer $n \geq k$, the bar W contains the point $[\xi]_n$. We hereby have to keep in mind that this procedure must be able to find such a k after the submission of *finitely* many points $[\xi]_1, [\xi]_2, \ldots, [\xi]_n, \ldots$ of the infinite sequence ξ – the notion one would need infinitely many data to start the calculation of k is nonsense.

To be precise: There must exist an assignment that appoints to each $x \in T$ after the submission of *finitely* many of the data $[x]_1, [x]_2, \ldots, [x]_n, \ldots$ a positive integer k – that of course depends on these finitely many data $[x]_1, [x]_2, \ldots, [x]_l$ – such that, for all points ξ fulfilling $[\xi]_1 = [x]_1, [\xi]_2 = [x]_2, \ldots, [\xi]_l = [x]_l$, and all integers $n \geq k$, the bar W contains the points $[\xi]_n$.

The procedure of detecting k therefore in principle must consist of the following two schemes:

In the *first scheme*, finitely many points $[x]_1, [x]_2, \ldots, [x]_j$ are submitted and the procedure is able to calculate from these data immediately a positive integer k such that, for all points ξ fulfilling $[\xi]_1 = [x]_1, [\xi]_2 = [x]_2, \ldots, [\xi]_j = [x]_j$, and all integers $n \geq k$, the bar W contains the points $[\xi]_n$.

In the *second scheme*, finitely many points $[x]_1, [x]_2, \ldots, [x]_j$ are submitted and the procedure requires the datum of the next point $[x]_{j+1}$ in order, either to calculate immediately a positive integer k such that, for all points ξ fulfilling $[\xi]_1 = [x]_1, [\xi]_2 = [x]_2, \ldots, [\xi]_j = [x]_j, [\xi]_{j+1} = [x]_{j+1}$, and all integers $n \geq k$, the bar W contains the points $[\xi]_n$, or to return to the starting-point of this second scheme.

We now show that *this second scheme can be eliminated:*

We consider the case that finitely many points $[x]_1, [x]_2, \ldots, [x]_j$ are submitted, and that the procedure – according to the second scheme – requires the datum of the next point $[x]_{j+1}$. The task that the procedure performs the following:

"*for each point $x \in S_{j+1}$ that fulfills for all $1 \leq j$ the inequality*

$$\|[x]_l - x\| \leq e_l + e_{j+1},$$

to calculate from the given data $[x]_1, [x]_2, \ldots, [x]_j$, x a positive integer k (that of course depends on x) such that, for all points ξ fulfilling $[\xi]_1 = [x]_1, [\xi]_2 = [x]_2, \ldots, [\xi]_j = [x]_j$, and $[\xi]_{j+1} = x$, and all integers $n \geq k$, the bar W contains the points $[\xi]_n$"

can be replaced by the device *just to announce the maximum of these integers k*. The justification for this replacement is as follows: The set S_{j+1} is finite. Thus we have only to take *finitely* many points x into account, with the consequence that only *finitely* many positive integers k have to be calculated. The maximum k' of these integers therefore is well-defined.

To sum up: We replace the procedure given so far by a modified version that starts with the *first* scheme given the data $[x]_1, [x]_2, \ldots, [x]_j$, i.e. that does not need, according to the second scheme, the indication of $x = [x]_{j+1}$, by immediately announcing this number k'. It can happen that this modified version discards possible smaller values of k for some of the possible points x, but the essential aim is achieved: The second scheme of the procedure, given the points $[x]_1, [x]_2, \ldots, [x]_j$, is eliminated.

We now think this process of elimination to be systematically accomplished at every occurrence of the second scheme in the original procedure or in the just adapted version of the procedure respectively, carrying with it a systematic revision of the procedure, with the effect that the announced positive integers k perhaps become gradually greater – but this does in no case damage the essential aim for which the procedure stands for.

Finally the revision of the procedure has eliminated all second schemes that appeared before. At this stage, the revised procedure starts as soon as a point $x \in S_1$ is given, and immediately announces a positive integer k^* such that, for all points ξ fulfilling $[\xi]_1 = x$, and all integers $n \geq k^*$, the bar W contains the points $[\xi]_n$. The set S_1 is finite. Therefore we have only to take *finitely* many points x into account, with the consequence that only *finitely* many positive integers k^* have to be calculated. The maximum of these integers is denoted by m, and it is evident that this integer m has the required property of the theorem. ∎

The most important consequence of the bar-theorem is the following corollary:

Theorem of HEINE **and** BOREL. *Let T denote a compact metric space, and suppose that one can construct for each point $\xi \in T$ a positive integer j such that a certain property $\mathcal{P}(\xi, l)$ holds for all integers $l \geq j$. Then it is possible to construct a positive integer k such that, for all points $\xi \in T$ and all integers $l \geq k$, the property $\mathcal{P}(\xi, l)$ holds.*

Proof. Let S denote a metric space with (Σ, E) as adequate pair of the two sequences

$$\Sigma = (S_1, S_2, \ldots, S_n, \ldots) \quad \text{and} \quad E = (e_1, e_2, \ldots, e_n, \ldots),$$

such that T is the completion of S, and all $S_1, S_2, \ldots, S_n, \ldots$ are finite. The assignment that picks up, given the point $\xi \in T$, the positive integer j such that $\mathcal{P}(\xi, l)$ holds for all integers $l \geq j$ must be a precisely defined *procedure*. If this procedure assigns the number j to the argument ξ, it must be able to start after the announcement of *finitely* many of the points

$$[\xi]_1, [\xi]_2, \ldots, [\xi]_n, \ldots$$

that define the argument ξ – the notion one would need infinitely many data to start the calculation of j is nonsense. In other words: The knowledge of a *sufficient* approximation $x = [\xi]_n$ of ξ is enough to fix this integer j.

Now the subset W of S is defined as follows: The fact that $x \in S_n$ belongs to W is equivalent to the possibility of calculating, given the datum $[\xi]_n = x$ for an arbitrary $\xi \in T$, the positive integer j such that $\mathcal{P}(\xi, l)$ holds for all integers $l \geq j$. The conclusion from above implies that W is a bar. We derive from the bar-theorem the existence of a positive integer m such that, for all points $\xi \in T$, at least $[\xi]_m$ belongs to W. Now it is possible to assign to each ξ, given the approximation $[\xi]_m \in S_m$, the positive integer j such that $\mathcal{P}(\xi, l)$ holds for all integers $l \geq j$. As S_m is finite, the maximum k of these integers j is a well-defined positive integer, and it guarantees, by construction, that, for all points $\xi \in T$ and all integers $l \geq k$, the property $\mathcal{P}(\xi, l)$ holds. ∎

The original version of the theorem of HEINE and BOREL does not rely upon the notion of a bar. This idea was introduced by BROUWER, and the proofs given here are essentially due to BROUWER, and are based on the *intuitionistic* concept of the nature of a sequence.

3.3 Topological concepts

3.3.1 The cover of a set

Let S denote a complete metric space, and let V denote a subspace of S. The set of all limit points of the metric space V is called the *cover* of V. The set V is called *closed* if and only if it coincides with its cover.

A point ξ belongs to the cover of V if and only if it is possible to find for each real $\varepsilon > 0$ a point $x \in V$ such that $\|x - \xi\| < \varepsilon$.

Proof. Suppose ξ belongs to the cover of V, i.e. ξ is a limit point of V. Then we have a convergent sequence X consisting of points $x_1, x_2, \ldots, x_n, \ldots$ belonging to V and fulfilling $\lim X = \xi$. It is possible to find for each real $\varepsilon > 0$ a positive integer j such that $n \geq j$ implies that $\|x_n - \xi\| < \varepsilon$. We therefore can set $x = x_j$. Suppose on the other hand that it is possible to find for each real $\varepsilon > 0$ a point $x \in V$ such that $\|x - \xi\| < \varepsilon$. Particularly it is possible to find for each positive integer n a point $x_n \in V$ such that $\|x_n - \xi\| < 10^{-n}$. The sequence X consisting of $x_1, x_2, \ldots, x_n, \ldots$ therefore represents a fundamental sequence of V with limit ξ, i.e. ξ belongs to the cover of V. ∎

As an example, let us consider the continuum as complete metric space with the absolute difference as its metric. Let j denote a positive integer. The set of all decimal numbers with exactly j decimal places is a closed set. The reason is as follows: If ξ belongs to the cover of this set, it must be possible to detect for each positive integer n a decimal number x_n with exactly j decimal places such that $|x_n - \xi| < 10^{-n}$, with the consequence $|x_n - x_j| \leq 10^{-n} + 10^{-j}$ for all positive integers n. This proves $x_n = x_j$ for all integers $n \geq j$ with the consequence that the sequence X consisting of $x_1, x_2, \ldots, x_n, \ldots$ has as its limit $\xi = \lim X = x_j$, i.e. ξ is a decimal number with exactly j decimal places.

If a set is totally bounded, its cover is totally bounded.

Proof. Let V denote a totally bounded set and $\varepsilon > 0$ an arbitrary real number. We construct a decimal number e such that $\varepsilon > e > 0$, and a finite $e/2$-net $(x_1, x_2, \ldots, x_n)$ of V. To each limit point ξ of V there exists a point $x \in V$ fulfilling $\|x - \xi\| < e/2$. It is further possible to detect a positive integer $j \leq n$ such that $\|x_j - x\| < e/2$. As this implies that $\|x_j - \xi\| \leq e < \varepsilon$, it is shown that $(x_1, x_2, \ldots, x_n)$ represents a finite ε-net of the cover of V. ∎

If a set is located, its cover is located.

Proof. Let V denote a located set, let ξ denote an arbitrary point, and let α, β denote arbitrary real numbers with $\alpha > \beta$. We construct two decimal numbers a, b fulfilling $\alpha > a > b > \beta$. Then at least one of the two cases,
1. there exists a $x_0 \in V$ such that $\|x_0 - \xi\| < \alpha$,
2. for all points $x \in V$ we have $\|x - \xi\| > a$,

must be true. If case 1 is true, then the point x_0 does not only belong to V, but also to its cover, and we are finished. Suppose in case 2 there would exist a limit point $\eta \in V$ such that $\|\eta - \xi\| < b$, then we could construct a point $y \in V$ such that $\|y - \eta\| < a - b$, with the consequence

$$\|y - \xi\| \leq b + (a - b) = a,$$

a contradiction to the assumption of case 2. Therefore in case 2 we have derived

$$\|\eta - \xi\| \geq b > \beta$$

for any limit point η of V, which proves the locatedness of the cover of V. ∎

3.3.2 The distance between a point and a set

Positive-distance-theorem. *Let V denote a located set and assume that the point η is apart from all points of the cover of V. Then we have $\inf \|V - \eta\| > 0$.*

Proof. For any positive integer n let e_n and d_n denote the decimal numbers

$$e_n = 10^{-n}, \quad d_n = e_n/2 = 5 \times 10^{-n-1}.$$

As V is supposed to be located at least one of the following two cases must take place:
Case 1: It is possible to construct a point $x_n \in V$ such that $\|x_n - \eta\| < e_n$.
Case 2: It is possible to prove $\|x - \eta\| > d_n$ for all $x \in V$.
We now define a procedure as follows: The procedure starts with an arbitrary point $x_0 \in V$. Let n denote a positive integer and suppose that the procedure has

already reached the point x_{n-1}. If, for this n, case 1 holds true, then x_n is defined to be a point of V such that $\|x_n - \eta\| < e_n$. If, for this n, case 2 holds true, then the procedure stops; we set $x_n = x_{n-1}$, and define further $x_{n+k} = x_n = x_{n-1}$ for all positive integers k.

We first prove for all positive integers n and m the inequalities

$$\|x_n - x_m\| \leq 2 \times 10^{-\min(n,m)} :$$

As long as the procedure runs, we have $\|x_n - \eta\| < e_n$ and $\|x_m - \eta\| < e_m$ together with

$$e_n + e_m = 10^{-n} + 10^{-m} \leq 2 \times 10^{-\min(n,m)} .$$

Suppose the procedure stops at the stage of the positive integer j: Then we derive for positive integers $n \leq j$ and $m \geq j$ the inequalities $\|x_n - x_j\| \leq e_n + e_j$ and $\|x_j - x_m\| = 0$, therefore $\|x_n - x_m\| \leq e_n + e_j$, which together with

$$e_n + e_j = 10^{-n} + 10^{-j} \leq 2 \times 10^{-n} = 2 \times 10^{-\min(n,m)}$$

again leads to the desired result. And if $n \geq j$ and $m \geq j$ is true we have the trivial situation $\|x_n - x_m\| = 0$.

Thus the sequence X consisting of the points $x_1, x_2, \ldots, x_n, \ldots$ is convergent, its limit $\xi = \lim X$ must belong to the cover of V and we even have the approximation $\|x_n - \xi\| \leq 2 \times 10^{-n}$ for all positive integers n. But η is apart from all points of the cover of V, in particular $\eta \neq \xi$, i.e. $\|\xi - \eta\| > 0$. There exists a positive integer k such that $\|\xi - \eta\| > 3 \times 10^{-k}$. For this k we derive from the triangle inequality

$$\begin{aligned}\|x_k - \eta\| &\geq \|\xi - \eta\| - \|\xi - x_k\| \geq \|\xi - \eta\| - 2 \times 10^{-k}\\ &\geq 3 \times 10^{-k} - 2 \times 10^{-k} = 10^{-k} = e_k .\end{aligned}$$

From this we see that at the latest at stage $n = k$ case 1 *cannot* be true. Therefore case 2 *must* hold true, telling us that we can prove $\|x - \eta\| > d_k$ for all points $x \in V$. This leads to the result $\inf \|V - \eta\| \geq d_k > 0$. ∎

3.3.3 The neighborhood of a point

Let S denote a complete metric space, and let ξ denote a point of S. For any real number $\alpha > 0$ the set $U_\alpha(\xi)$ of all points x of the metric space fulfilling $\|\xi - x\| < \alpha$ is called the *α-neighborhood of ξ*.

If a point x belongs to the cover of the α-neighborhood $U_\alpha(\xi)$ of the point ξ, then we have $\|\xi - x\| \leq \alpha$.

Proof. If x belongs to the cover of the α-neighborhood $U_\alpha(\xi)$ of the point ξ, then x is a limit point of a sequence X consisting of points $x_1, x_2, \ldots, x_n, \ldots$ that belong to $U_\alpha(\xi)$, i.e. we have for all positive integers n the inequalities

$$\|\xi - x_n\| < \alpha .$$

Therefore, by the principle of permanence, $\|\xi - x\| = \lim \|\xi - X\| \leq \alpha$. ∎

Let S denote a complete metric space, and let V denote a subspace of S. A point ξ is called an *inner point* of V if and only if it is possible to detach a real $\delta > 0$ such that all points of the δ-neighborhood $U_\delta(\xi)$ belong to V. Suppose that inner points of V exist; then the set consisting of all inner points of the metric space V is called the *interior* of V. The set V is called *open* if and only if it coincides with its interior.

Let S denote a complete metric space, and let V denote a subspace of S. A point ξ is called an *outer point* of V if it is possible to detach a real $\delta > 0$ such that all points of the δ-neighborhood $U_\delta(\xi)$ are apart from any point belonging to V. Suppose that outer points of V exist; then the set consisting of all outer points of the set V is called the *exterior* of V.

Suppose for a point ξ and a real $\delta > 0$ that the set V comprises $U_\delta(\xi)$. Then all points of $U_\delta(\xi)$ are inner points of V.

Proof. Let η denote an arbitrary point of $U_\delta(\xi)$. It is possible to construct a decimal number d such that

$$\delta - \|\xi - \eta\| > d > 0.$$

For any point x of the d-neighborhood $U_d(\eta)$ of η we have $\|\eta - x\| < d$, and therefore

$$\|\xi - x\| - \|\xi - \eta\| \leq \|x - \eta\| < d < \delta - \|\xi - \eta\|.$$

This shows $\|\xi - x\| < \delta$, i.e. x is contained in $U_\delta(\xi)$, and a fortiori contained in V. Thus η proves to be an inner point of V. ∎

Each α-neighborhood $U_\alpha(\xi)$ of a point ξ is an open set.

Proof. Identify V with $U_\alpha(\xi)$ in the proposition above. ∎

Let S denote a complete metric space, and let V denote a subspace of S such that outer points of V exist. Then the exterior of V is an open set.

Proof. Let ξ denote a point of the exterior of V. It is possible to detach a real $\delta > 0$ such that all points of the δ-neighborhood $U_\delta(\xi)$ are apart from any point belonging to V. Let η denote an arbitrary point of $U_\delta(\xi)$. It is possible to construct a decimal number d such that

$$\delta - \|\xi - \eta\| > d > 0.$$

For any point x of the d-neighborhood $U_d(\eta)$ of η we have $\|\eta - x\| < d$, and therefore

$$\|\xi - x\| - \|\xi - \eta\| \leq \|x - \eta\| < d < \delta - \|\xi - \eta\|.$$

This shows $\|\xi - x\| < \delta$, i.e. x is contained in $U_\delta(\xi)$, and a fortiori apart from any point belonging to V. Thus η proves to be an outer point of V, and ξ proves to be an inner point of the exterior of V. ∎

Let S denote a complete metric space, and let V denote a subspace of S such that outer points of V exist. Then each point of the exterior of V is apart from any point of V. If V is a located and closed set, the converse is also true: Each point that is apart from any point of V belongs to the exterior of V.

Proof. The first part of the proposition is obvious as each point ξ belongs to its own δ-neighborhood $U_\delta(\xi)$.

To prove the second part we derive from the positive-distance-theorem: for any point ξ that is apart from any point of V we have $\inf \|V - \xi\| = \delta > 0$. Suppose η is an arbitrary point belonging to $U_\delta(\xi)$. We have, for each point x in V, the chain of inequalities

$$\|x - \eta\| \geq \|x - \xi\| - \|\eta - \xi\| > \delta - \delta = 0,$$

and this proves that η is apart from any point of V. ∎

3.3.4 Dense and nowhere dense

Let S denote a complete metric space, and let V denote a subspace of S. V is called *dense* if and only if for each point $\xi \in S$ and each real $\varepsilon > 0$ it is possible to detach a point of V belonging to the ε-neighborhood $U_\varepsilon(\xi)$ of ξ. (A *sequence* is called dense if and only if the set consisting of all its points is dense.)

Let S denote a complete metric space, and let V denote a subspace of S. V is dense if and only if the cover of V coincides with S.

Proof. We already know that a point ξ belongs to the cover of V if and only if it is possible to find, for each real $\varepsilon > 0$, a point $x \in V$ such that $\|x - \xi\| < \varepsilon$, i.e. x belongs to $U_\varepsilon(\xi)$. ∎

Let S denote a complete metric space, and let V denote a subspace of S. V is called *nowhere dense* if and only if for each point $\xi \in S$ and each real $\varepsilon > 0$ it is possible to detach an outer point of V belonging to the ε-neighborhood $U_\varepsilon(\xi)$ of ξ. (A *sequence* is called nowhere dense if and only if the set consisting of all its points is nowhere dense.)

Theorem of BAIRE. *Let $(V_1, V_2, \ldots V_n, \ldots)$ denote a sequence of located and nowhere dense sets in the complete metric space S. Then for any point $x_0 \in S$ and any real $\varepsilon > 0$ it is possible to detach a point ξ belonging to the ε-neighborhood $U_\varepsilon(x_0)$ of x_0 which is an outer point of each of the sets $V_1, V_2, \ldots, V_n, \ldots$.*

Proof. We define a procedure that starts with the point x_0 and a decimal number d_0 such that $\varepsilon > d_0 > 0$. The procedure will define step by step points $x_1, x_2, \ldots, x_n, \ldots$ and positive decimal numbers $d_1, d_2, \ldots, d_n, \ldots$ in the following way: Suppose we already know the point x_{n-1} and the positive decimal number d_{n-1}. As V_n is nowhere dense it is possible to detach an outer point $x_n \in U_{d_{n-1}}(x_{n-1})$. It is further possible to construct a positive decimal number d'_n such that all points belonging to $U_{d'_n}(x_n)$ are outer points of V_n. The inequality $\|x_n - x_{n-1}\| < d_{n-1}$

and the interpolation lemma finally allow us to construct two positive decimal numbers d_n''' and c_n such that we even have

$$\|x_n - x_{n-1}\| < c_n < d_{n-1} - d_n''.$$

By doing this we conclude for all points $x \in U_{d_n''}(x_n)$ from the two inequalities

$$\|x - x_n\| < d_n'' \quad \text{and} \quad \|x_n - x_{n-1}\| < c_n$$

the relation

$$\|x - x_{n-1}\| \leq d_n'' + c_n < d_n'' + (d_{n-1} - d_n'') = d_{n-1},$$

i.e. all these points x belong to the d_{n-1}-neighborhood $U_{d_{n-1}}(x_{n-1})$ of x_{n-1}. We now define d_n to be a decimal number with the property

$$0 < d_n < \min\left(d_n', d_n'', 10^{-n}\right).$$

This has the following three consequences:
First: the cover of $U_{d_n}(x_n)$ is contained in $U_{d_n'}(x_n)$ which proves that all points of this cover are outer points of V_n.
Second: the cover of $U_{d_n}(x_n)$ is contained in $U_{d_n''}(x_n)$ and therefore contained in the open set $U_{d_{n-1}}(x_{n-1})$. A fortiori the open set $U_{d_n}(x_n)$ itself is contained in the open set $U_{d_{n-1}}(x_{n-1})$.
Third: We have for all positive integers n and k the inequality

$$\|x_n - x_{n+k}\| \leq d_n \leq 10^{-n}$$

which shows that the sequence X consisting of $x_1, x_2, \ldots, x_n, \ldots$ converges. The limit $\xi = \lim X$ fulfills for each positive integer n the inequality

$$\|x_n - \xi\| \leq d_n.$$

Therefore ξ belongs for each positive integer n to the cover of $U_{d_n}(x_n)$, and by the first consequence has to be an outer point of V_n. Further, by the second consequence, ξ belongs to $U_{d_{n-1}}(x_{n-1})$ and a fortiori to the ε-neighborhood $U_\varepsilon(x_0)$ of x_0. ∎

Theorem of CANTOR. *For any real number α, for any real $\varepsilon > 0$, and for any infinite sequence C consisting of real numbers $\gamma_1, \gamma_2, \ldots, \gamma_n, \ldots$ there exists a real number β with $|\alpha - \beta| < \varepsilon$ such that β is apart from each of the numbers $\gamma_1, \gamma_2, \ldots, \gamma_n, \ldots$.*

Proof. Let S_n denote the space of all decimal numbers a of the form

$$a = z + 0.z_1 z_2 \ldots z_n = z + z_1 \times 10^{-1} + z_2 \times 10^{-2} + \ldots + z_n \times 10^{-n},$$

z being an integer and $z_1, z_2, \ldots, z_n$ being integers with $0 \leq z_j \leq 9$ for all positive integers $j \leq n$. Define S as the union of $S_1, S_2, \ldots, S_n, \ldots$, i.e. the set of all decimal numbers. The metric on S is the absolute difference and the sequence E consists of the numbers $e_n = 10^{-n}$ for all positive integers n. The completion T of S is the continuum.

For any positive integer n the set V_n only consisting of γ_n is located and nowhere dense. Then it is possible, by BAIRE's theorem, for any point $\alpha \in T$ and any real $\varepsilon > 0$, to detach a point β belonging to the ε-neighborhood $U_\varepsilon(\alpha)$ of α which is outer point of each of the sets $V_1, V_2, \ldots, V_n, \ldots$. ∎

3.3.5 Connectedness

Let $\alpha > 0$ denote a real number. A finite sequence X of points $x_0, x_1, x_2, \ldots, x_j$ is called an α-*connection* if and only if for all positive integers $n \leq j$ the inequalities $\|x_{n-1} - x_n\| < \alpha$ hold. The α-connection X is called an α-*cycle* if and only if we have furthermore $\|x_j - x_0\| < \alpha$.

Let S denote a complete metric space, and let V denote a subspace of S. We say that V *connects* two points ξ and η of S if and only if it is possible for any real $\varepsilon > 0$ to construct an ε-connection X of points $x_0, x_1, x_2, \ldots, x_j \in V$ such that the inequalities $\|\xi - x_0\| < \varepsilon$ and $\|x_j - \eta\| < \varepsilon$ hold.

If V connects the two points ξ and η, these points have to belong to the cover of V.

Proof. Suppose that V connects the two points ξ and η. Then it is possible for any positive integer n to construct a 10^{-n}-connection X_n of points $x_0^{(n)}, x_1^{(n)}, x_2^{(n)}, \ldots, x_{j_n}^{(n)} \in V$ such that the inequalities $\left\|\xi - x_0^{(n)}\right\| < 10^{-n}$ and $\left\|x_{j_n}^{(n)} - \eta\right\| < 10^{-n}$ hold. Particularly the two sequences

$$X^* = \left(x_0^{(1)}, x_0^{(2)}, \ldots, x_0^{(n)} \ldots\right)$$

and

$$X^{**} = \left(x_{j_1}^{(1)}, x_{j_2}^{(2)}, \ldots, x_{j_n}^{(n)}, \ldots\right)$$

of points belonging to V are convergent with $\lim X^* = \xi$ and $\lim X^{**} = \eta$. ∎

V connects the two points ξ and η if and only if the cover of V connects these points.

Proof. If V connects the two points ξ and η, then it is obvious that the cover of V connects these points.

Now suppose that the cover of V connects the points ξ and η and let $\varepsilon > 0$ denote an arbitrary real number. We construct a decimal number e such that $\varepsilon > e > 0$ and construct an $(e/2)$-connection X of points $\xi_0, \xi_1, \xi_2, \ldots, \xi_j$ belonging to the cover of V such that the inequalities $\|\xi - \xi_0\| < e/2$ and $\|\xi_j - \eta\| < e/2$ hold. For each integer n with $0 \leq n \leq j$ it is possible to construct a point $x_n \in V$ such that $\|\xi_n - x_n\| < e/4$. For all positive integers $n \leq j$ the three inequalities

$$\|x_{n-1} - \xi_{n-1}\| < \frac{e}{4}$$
$$\|\xi_{n-1} - \xi_n\| < \frac{e}{2}$$
$$\|\xi_n - x_n\| < \frac{e}{4}$$

lead to $\|x_{n-1} - x_n\| \leq e < \varepsilon$. ∎

3.4 The s-dimensional continuum

3.4.1 Metrics in the s-dimensional space

Let s and n denote positive integers. $\mathbb{D}_n^s$ is the set of all s-tupels $x = (a_1, \ldots, a_s)$, the *points* of $\mathbb{D}_n^s$, the *coordinates* $a_1, \ldots, a_s$ being decimal numbers with exactly n decimal places. $\mathbb{D}^s$ is the set of all s-tupels $x = (a_1, \ldots, a_s)$, the *points* of $\mathbb{D}^s$, the *coordinates* $a_1, \ldots, a_s$ being decimal numbers.

$\mathbb{D}^s$ *is a metric space: for any two points* $x = (a_1, \ldots, a_s)$, $y = (b_1, \ldots, b_s)$ *a distance* $\|x - y\|_\infty$ *between x and y can be defined as*

$$\|x - y\|_\infty = \max(|a_1 - b_1|, \ldots, |a_s - b_s|).$$

Proof. The positivity, i.e. $\|x - y\|_\infty \geq 0$ for all points x, y, is obvious. Suppose $x = (a_1, \ldots, a_s) \neq y = (b_1, \ldots, b_s)$. This is equivalent to the fact that it is possible to detect a positive integer $j \leq s$ such that $a_j \neq b_j$. This implies that

$$\|x - y\|_\infty = \max(|a_1 - b_1|, \ldots, |a_s - b_s|) \geq |a_j - b_j| > 0.$$

Suppose on the other hand

$$\|x - y\|_\infty = \max(|a_1 - b_1|, \ldots, |a_s - b_s|) > 0.$$

Then at least one of the absolute differences $|a_1 - b_1|, \ldots, |a_s - b_s|$ has to be positive, i.e. it is possible to detect a positive integer $j \leq s$ such that $a_j \neq b_j$. This implies that $x \neq y$. Thus this distance proves to be definite.
The triangle inequality follows with the notations

$$x = (a_1, \ldots, a_s), \quad y = (b_1, \ldots, b_s), \quad z = (c_1, \ldots, c_s)$$

from the calculation

$$\begin{aligned}
\|z - x\|_\infty &= \max(|a_1 - c_1|, \ldots, |a_s - c_s|) \\
&\leq \max(|a_1 - b_1| + |b_1 - c_1|, \ldots, |a_s - b_s| + |b_s - c_s|) \\
&\leq \max(|a_1 - b_1|, \ldots, |a_s - b_s|) \\
&\quad + \max(|c_1 - b_1|, \ldots, |c_s - b_s|) \\
&= \|x - y\|_\infty + \|z - y\|_\infty
\end{aligned}$$

which implies that
$$\|z-x\|_\infty - \|z-y\|_\infty \leq \|x-y\|_\infty .$$

∎

A metric, i.e. a distance $\|x-y\|$ between the two points $x, y \in \mathbb{D}^s$, is called *equivalent* to the metric given above if it is possible to find two positive decimal numbers c' and c'' with the property that we have for all points x, y the chain of inequalities
$$c' \|x-y\|_\infty \leq \|x-y\| \leq c'' \|x-y\|_\infty .$$

All concepts and all perceptions we had about metric spaces so far do not change if the given metric is replaced by an equivalent metric.

For example: Given any two points $x = (a_1, \ldots, a_s)$, $y = (b_1, \ldots, b_s)$, a distance $\|x-y\|_1$ between x and y can be defined as
$$\|x-y\|_1 = |a_1 - b_1| + \ldots + |a_s - b_s| .$$

It is equivalent to the metric defined above, because we have
$$\|x-y\|_\infty \leq \|x-y\|_1 \leq s \|x-y\|_\infty .$$

3.4.2 The completion of the s-dimensional space

The completion of $\mathbb{D}^s$ is called the *s-dimensional continuum* and denoted by $\mathbb{R}^s$.

Each point $\xi \in \mathbb{R}^s$ can be written as $\xi = (\alpha_1, \ldots, \alpha_s)$ with real coordinates $\alpha_1, \ldots, \alpha_s$. Conversely each s-tupel $(\alpha_1, \ldots, \alpha_s)$ with real coordinates $\alpha_1, \ldots, \alpha_s$ can be identified with a point in $\mathbb{R}^s$. We abbreviate $\mathbb{D}^1 = \mathbb{D}$ and $\mathbb{R}^1 = \mathbb{R}$.

Proof. The pair (Σ, E) consisting of the two sequences
$$\Sigma = \left(\mathbb{D}_1^s, \mathbb{D}_2^s, \ldots, \mathbb{D}_n^s, \ldots\right) \quad \text{and} \quad E = \left(10^{-1}, 10^{-2}, \ldots, 10^{-n}, \ldots\right)$$
is an adequate pair in the metric space $\mathbb{D}^s$.

Let ξ denote a limit point, i.e. a sequence consisting of points
$$[\xi]_1, [\xi]_2, \ldots, [\xi]_n, \ldots$$
belonging to $\mathbb{D}_1^s, \mathbb{D}_2^s, \ldots, \mathbb{D}_n^s, \ldots$ respectively with
$$[\xi]_n = \left(a_1^{(n)}, \ldots, a_s^{(n)}\right) .$$

For any positive integers n and m we have
$$\left\|[\xi]_n - [\xi]_m\right\|_\infty = \max\left(\left|a_1^{(n)} - a_1^{(m)}\right|, \ldots, \left|a_s^{(n)} - a_s^{(m)}\right|\right) \leq 10^{-n} + 10^{-m} .$$

This is accompanied by the existence of s sequences $A_1, \ldots, A_s$: for all integers $j \leq s$ the sequence A_j consists of the j-th coordinates $a_j^{(1)}, a_j^{(2)}, \ldots, a_j^{(n)}, \ldots$ of $[\xi]_1, [\xi]_2, \ldots, [\xi]_n, \ldots$. We have, for each positive integer $j \leq s$ and for all positive integers n, m, the inequalities

$$\left| a_j^{(n)} - a_j^{(m)} \right| \leq 10^{-n} + 10^{-m},$$

and this implies that each of the s sequences $A_1, \ldots, A_s$ is convergent. Therefore the coordinates

$$\alpha_1 = \lim A_1, \quad \ldots, \quad \alpha_s = \lim A_s$$

of ξ are well-defined.

Suppose, on the contrary, we have an s-tupel $\xi = (\alpha_1, \ldots, \alpha_s)$ of real numbers $\alpha_1, \ldots, \alpha_s$. We then define a sequence X of points $x_1, x_2, \ldots, x_n, \ldots$ by

$$x_n = ([\alpha_1]_n, \ldots, [\alpha_s]_n).$$

As for each positive integer n the point x_n belongs to $\mathbb{D}_n^s$ and the inequalities

$$\begin{aligned}
\|x_n - x_m\|_\infty &= \max(|[\alpha_1]_n - [\alpha_1]_m|, \ldots, |[\alpha_s]_n - [\alpha_s]_m|) \\
&\leq 10^{-n} + 10^{-m}
\end{aligned}$$

hold for all positive integers n and m, the point $\xi = \lim X$ in the complete metric space $\mathbb{R}^s$ proves to be a limit point with the given real numbers $\alpha_1, \ldots, \alpha_s$ as coordinates. ∎

Two points $\xi = (\alpha_1, \ldots, \alpha_s)$ and $\eta = (\beta_1, \ldots, \beta_s)$ of the s-dimensional continuum are apart if and only if it is possible to detach a positive integer $j \leq s$ such that $\alpha_j \neq \beta_j$. They are equal if and only if their corresponding coordinates are equal.

Proof. The second part of the proposition immediately follows from the first part. Suppose that ξ and η are apart, i.e. $\|\xi - \eta\|_\infty > 0$. The real number $\|\xi - \eta\|_\infty$ is the limit of the sequence of the real numbers $\|[\xi]_1 - [\eta]_1\|_\infty, \|[\xi]_2 - [\eta]_2\|_\infty,$ $\ldots, \|[\xi]_n - [\eta]_n\|_\infty, \ldots$, and we assume that, for all positive integers n, the points $[\xi]_n$ and $[\eta]_n$ are given as

$$[\xi]_n = \left(a_1^{(n)}, \ldots, a_s^{(n)}\right), \quad [\eta]_n = \left(b_1^{(n)}, \ldots, b_s^{(n)}\right).$$

For all positive integers n and for all positive integers $j \leq s$ we have

$$\left| a_j^{(n)} - \alpha_j \right| \leq 10^{-n} \quad \text{and} \quad \left| b_j^{(n)} - \beta_j \right| \leq 10^{-n}.$$

We construct a decimal number d with $\|\xi - \eta\|_\infty > d > 0$ and a positive integer k such that for all $n \geq k$

$$\left| \|[\xi]_n - [\eta]_n\|_\infty - \|\xi - \eta\|_\infty \right| < \frac{d}{4}$$

holds. Let n denote a positive integer with $n \geq k$ and $10^{-n} \leq d/4$. Then we can pick up a positive integer $j \leq s$ fulfilling

$$\left|a_j^{(n)} - b_j^{(n)}\right| > \frac{3d}{4}.$$

Because if this were not the case, we would have $\|[\xi]_n - [\eta]_n\|_\infty \leq 3d/4$ together with $\|\xi - \eta\|_\infty > d$ and thus the contradiction

$$\|\xi - \eta\|_\infty - \|[\xi]_n - [\eta]_n\|_\infty > d - \frac{3d}{4} = \frac{d}{4}$$

to the inequality

$$\left|\|\xi - \eta\|_\infty - \|[\xi]_n - [\eta]_n\|_\infty\right| < \frac{d}{4}.$$

If we had $|\alpha_j - \beta_j| < d/4$, the three inequalities

$$\left|a_j^{(n)} - \alpha_j\right| \leq \frac{d}{4}, \quad |\alpha_j - \beta_j| < \frac{d}{4}, \quad \left|\beta_j - b_j^{(n)}\right| \leq \frac{d}{4}$$

implied the contradiction

$$\left|a_j^{(n)} - b_j^{(n)}\right| \leq \frac{3d}{4}$$

to the relation $\left|a_j^{(n)} - b_j^{(n)}\right| > 3d/4$. Therefore we have

$$|\alpha_j - \beta_j| \geq \frac{d}{4} > 0,$$

in particular $\alpha_j \neq \beta_j$.

Suppose on the other hand $\xi = (\alpha_1, \ldots, \alpha_s)$, $\eta = (\beta_1, \ldots, \beta_s)$, and the existence of a positive integer $j \leq s$ such that $\alpha_j \neq \beta_j$. We write for all integers n

$$[\xi]_n = \left(a_1^{(n)}, \ldots, a_s^{(n)}\right), \quad [\eta]_n = \left(b_1^{(n)}, \ldots, b_s^{(n)}\right)$$

and derive from the principle of permanence and the inequalities

$$\|[\xi]_n - [\eta]_n\|_\infty = \max\left(\left|a_1^{(n)} - b_1^{(n)}\right|, \ldots \left|a_s^{(n)} - b_s^{(n)}\right|\right)$$
$$\geq \left|a_j^{(n)} - b_j^{(n)}\right|$$

the inequality

$$\|\xi - \eta\|_\infty \geq |\alpha_j - \beta_j| > 0.$$

Therefore $\xi \neq \eta$ must be true. ∎

3.4.3 Cells, rays, and linear subspaces

Two points $\xi = (\alpha_1, \ldots, \alpha_s)$ and $\eta = (\beta_1, \ldots, \beta_s)$ of $\mathbb{R}^s$ are called *discrete* if and only if it is possible, for each positive integer $j \leq s$, to decide whether $\alpha_j \neq \beta_j$ or $\alpha_j = \beta_j$ is true, Hence it is possible to divide the set of the positive integers $j \leq s$ into two disjoint subsets J' and J'': $j \in J'$ implies that $\alpha_j \neq \beta_j$, and $j \in J''$ implies that $\alpha_j = \beta_j$.

Let $\xi = (\alpha_1, \ldots, \alpha_s)$ and $\eta = (\beta_1, \ldots, \beta_s)$ denote two discrete points, assume that the set of the positive integers $j \leq s$ is divided into two disjoint subsets J' and J'': $j \in J'$ implies that $\alpha_j \neq \beta_j$, and $j \in J''$ implies that $\alpha_j = \beta_j$. Assume further that for all $j \in J'$ the inequality $\alpha_j < \beta_j$ holds.
Then the set $[\xi; \eta]$ of all points $x = (x_1, \ldots, x_s)$ with $\alpha_j \leq x_j \leq \beta_j$ for all positive integers $j \leq s$ is called a *cell*. The number k of elements of the set J' is called the *dimension* of the cell.
The cell $[\xi; \eta]$ is embedded in two k-dimensional *rays* $[\xi; \eta\rangle$ and $\langle\xi; \eta]$: The first consists of all points $x = (x_1, \ldots, x_s)$ with $\alpha_j \leq x_j$ for all $j \in J'$, and with $\alpha_j = x_j = \beta_j$ for all $j \in J''$. The second consists of all points $x = (x_1, \ldots, x_s)$ with $x_j \leq \beta_j$ for all $j \in J'$, and with $\alpha_j = x_j = \beta_j$ for all $j \in J''$.
These two rays finally are embedded in a k-dimensional *linear subspace* $\langle\xi; \eta\rangle$ that is *spanned* by ξ and η: this subspace consists of all points $x = (x_1, \ldots, x_s)$ with $\alpha_j = x_j = \beta_j$ for all $j \in J''$.

Cells, rays, and linear subspaces are closed sets.

Proof. This is an consequence of the principle of permanence. ∎

Let $\xi = (\alpha_1, \ldots, \alpha_s)$ and $\eta = (\beta_1, \ldots, \beta_s)$ denote two discrete points, assume that the set of the positive integers $j \leq s$ is divided into two disjoint subsets J' and J'': $j \in J'$ implies that $\alpha_j \neq \beta_j$, and $j \in J''$ implies that $\alpha_j = \beta_j$. Assume further that for all $j \in J'$ the inequality $\alpha_j < \beta_j$ holds and that J' contains at least one number.
Then $]\xi; \eta[$ or $[\xi; \eta[$ or $]\xi; \eta]$ are defined as sets of all points $x = (x_1, \ldots, x_s)$ with $\alpha_j < x_j < \beta_j$ or $\alpha_j \leq x_j < \beta_j$ or $\alpha_j < x_j \leq \beta_j$ respectively for all $j \in J'$, and with $\alpha_j = x_j = \beta_j$ for all $j \in J''$.
$]\xi; \eta\rangle$ is defined to consists of all points $x = (x_1, \ldots, x_s)$ with $\alpha_j < x_j$, for all $j \in J'$, and with $\alpha_j = x_j = \beta_j$ for all $j \in J''$. $\langle\xi; \eta[$ is defined to consists of all points $x = (x_1, \ldots, x_s)$ with $x_j < \beta_j$ for all $j \in J'$, and with $\alpha_j = x_j = \beta_j$ for all $j \in J''$.

The sets $]\xi; \eta[,]\xi; \eta\rangle, \langle\xi; \eta[$ are open sets in the metric space $\langle\xi; \eta\rangle$.

Proof. Suppose we have the inequality $\alpha_j < \gamma_j$ or $\gamma_j < \beta_j$ for an integer $j \in J'$. Then there exists a positive decimal number d such that $d < \gamma_j - \alpha_j$ or $d < \beta_j - \gamma_j$. But this proves for all x_j fulfilling $|x_j - \gamma_j| < d$ the relation $\alpha_j < x_j$ or $x_j < \beta_j$. ∎

In the special case $s = 1$ and $\alpha < \beta$ the sets $[\alpha; \beta]$ and $]\alpha; \beta[$ are called *closed and open bounded intervals*, the sets $[\alpha; \beta[$ and $]\alpha; \beta]$ are called *half-open intervals*. The sets $[\alpha; \beta\rangle$ and $\langle\alpha; \beta]$ are denoted as $[\alpha; \infty[$ and $]\infty; \beta]$ and are called *closed unbounded intervals*. The sets $]α; β\rangle$ and $\langle α; β[$ are denoted as $]α; ∞[$ and $]∞; β[$ and are called *open unbounded intervals*.

3.4.4 Totally bounded sets in the s-dimensional continuum

Theorem of WEIERSTRASS. *A set in the s-dimensional continuum $\mathbb{R}^s$ is totally bounded if and only if it is bounded and located.*

Proof. It is clear that a totally bounded set is bounded and located.
Suppose on the other hand that a set in the s-dimensional continuum is bounded and located. Then it suffices to prove that it is the subset of a compact metric space. This is true because the set can be embedded into a cell of the form $[ξ; η]$ with
$$ξ = (-k, -k, \ldots, -k) \quad \text{and} \quad η = (k, k, \ldots, k),$$
k denoting a sufficiently big positive integer.
Define for any positive integer n the set S_n to consist of the points $x = (a_1, \ldots, a_s)$ where for all positive integers $j \leq s$ the decimal numbers a_j are given as
$$a_j = z + w_1 \times 10^{-1} + w_2 \times 10^{-2} + \ldots + w_n \times 10^{-n},$$
$z, w_1, w_2, \ldots, w_n$ being integers with
$$-k < z < k, \quad -9 \leq w_1 \leq 9, \quad -9 \leq w_2 \leq 9, \quad \ldots, \quad -9 \leq w_n \leq 9.$$
It is clear that (Σ, E) with
$$\Sigma = (S_1, S_2, \ldots, S_n, \ldots) \quad \text{and} \quad E = \left(10^{-1}, 10^{-2}, \ldots, 10^{-2}, \ldots\right)$$
forms an adequate pair with $[ξ; η]$ as corresponding complete metric space. As all S_n are finite, $[ξ; η]$ must be compact. ∎

3.4.5 The supremum and the infimum

Suppose the set U of the continuum is located and bounded. Then there exists a real number $σ$, the so-called supremum $σ = \sup U$ of the set U, with the following two properties:
$σ$ is an upper bound of U, i.e. we have $x \leq σ$ for all $x \in U$,
$σ$ is the smallest upper bound of U, i.e. for all real numbers $α < σ$ it is possible to find a point $x_0 \in U$ such that $x_0 > α$.

Proof. As U is bounded, there exists a decimal number c such that we have $x \leq c$ for all $x \in U$. This implies that
$$|x - c| = c - x$$
and we can calculate the infimum
$$μ = \inf |U - c|.$$

We now prove that $\sigma = c - \mu$ has the required properties:
By definition of μ we have

$$|x - c| = c - x \geq \mu$$

for all $x \in U$, and this implies that

$$x \leq c - \mu = \sigma \,.$$

Suppose $\alpha < \sigma$, then we have by setting $\lambda = c - \alpha$ the relation

$$\lambda = c - \alpha > c - \sigma = \mu \,,$$

and this implies, by the definition of μ, the existence of a real number $x_0 \in U$ such that

$$|x_0 - c| = c - x_0 < \lambda \,,$$

which leads to $x_0 > c - \lambda = \alpha$. ∎

Suppose the set U of the continuum is located and bounded. Then there exists a real number ρ, the so-called infimum $\rho = \inf U$ of the set U, with the following two properties:
ρ is a lower bound of U, i.e. we have $x \geq \rho$ for all $x \in U$,
ρ is the greatest lower bound of U, i.e. for all real numbers $\alpha > \rho$ it is possible to find a point $x_0 \in U$ such that $x_0 < \alpha$.

Proof. As U is bounded, there exists a decimal number c such that we have

$$0 - x \leq c$$

for all $x \in U$. We now define V to be the set of all $y = 0 - x$ with $x \in U$. We, by the foregoing proposition, can calculate $\tau = \sup V$ and define $\rho = 0 - \tau$. We now prove that ρ has the required properties:
First by definition we have $0 - x \leq \tau$ for all $x \in U$. This implies that

$$x \geq 0 - \tau = \rho \,.$$

Suppose $\alpha > \rho$, then the real number $\beta = 0 - \alpha$ fulfills the inequality

$$\beta = 0 - \alpha < 0 - \rho = \tau$$

from which, by definition of τ, the existence of a real number $x_0 \in U$ with

$$0 - x_0 > \beta$$

follows. This implies that

$$x_0 < 0 - \beta = \alpha \,.$$

∎

3.4.6 Compact intervals

A compact interval is a connected set in the continuum.

Proof. α, β denote two real numbers with $\alpha < \beta$, and V denotes the set of all decimal numbers x fulfilling $\alpha \leq x \leq \beta$. The compact interval $[\alpha; \beta]$ is the cover of V. Let $\varepsilon > 0$ be an arbitrary real number. Suppose x' and x'' denote two decimal numbers of V of the form

$$x' = p' \times 10^{-n}, \quad x'' = p'' \times 10^{-n},$$

p', p'' being integers; the positive integer n can be chosen so big that $10^{-n} < \varepsilon$. We further can assume without loss of generality that $p' \leq p''$. Then the finite sequence consisting of all

$$x_j = (p' + j) \times 10^{-n}, \quad 0 \leq j \leq p'' - p'$$

represents an ε-connection that connects x' with x''. ∎

Suppose a compact and connected subset of the continuum consists of at least two different points. Then this subset is a compact interval.

Proof. Denote this subset by T. We can calculate the real numbers $\alpha = \inf T$ and $\beta = \sup T$. Two real numbers ξ, η with $\xi < \eta$ are members of T and we therefore have

$$\alpha \leq \xi < \eta \leq \beta.$$

This implies that $\alpha < \beta$ and that T is subset of the compact interval $[\alpha; \beta]$.
Let c denote an arbitrary decimal number in $[\alpha; \beta]$ and $\varepsilon > 0$ an arbitrary real number. We construct a decimal number e with $\varepsilon > e > 0$. T connects the real numbers α and β, therefore the construction of a finite sequence $(\xi_0, \xi_1, \ldots, \xi_j)$ of real numbers in T can be performed with the properties

$$|\alpha - \xi_0| < e, \quad |\xi_n - \beta| < e,$$

and $|\xi_{n-1} - \xi_n| < e$ for all positive integers $n \leq j$.
Suppose we had $|\xi_n - c| > e$ for all integers n with $0 \leq n \leq j$. The following argument shows that this supposition is absurd:
It would lead to the consequence that we could divide the set of integers n with $0 \leq n \leq j$ into two disjoint sets J' and J'' in the following way: $n \in J'$ would be equivalent to $c - \xi_n > e$, i.e. $\xi_n < c - e$, and $n \in J''$ would be equivalent to $\xi_n - c > e$, i.e. $\xi_n > c + e$. The relations

$$|\alpha - \xi_0| = \xi_0 - \alpha < e \quad \text{and} \quad \alpha \leq c$$

imply $\xi_0 - c < e$ and thus $0 \in J'$. The relations

$$|\xi_j - \beta| = \beta - \xi_j < e \quad \text{and} \quad c \leq \beta$$

imply $c - \xi_j < e$ and thus $j \in J''$. J'' therefore has at least one element, and we can pick up the *smallest* integer n that lies in J''. As 0 is an element of J', we have $n > 0$, and the integer $n - 1$ must be an element of J'. The two inequalities

$$\xi_n > c + e \quad \text{and} \quad \xi_{n-1} < c - e$$

then would imply

$$\xi_n - \xi_{n-1} > \xi_n - (c - e) > (c + e) - (c - e) = 2e,$$

a relation in contradiction to $|\xi_n - \xi_{n-1}| < e$.

Thus we have proved that there must exist a real number ξ_n belonging to the finite sequence $(\xi_0, \xi_1, \ldots, \xi_j)$ such that $|c - \xi_n| \leq e$. In other words: To each decimal number $c \in [\alpha; \beta]$ and to each real $\varepsilon > 0$ there exists a real number $\xi \in T$ such that $|c - \xi| < \varepsilon$. T is a closed set, therefore we derive that c itself belongs to T.

Finally each real number $\gamma \in [\alpha; \beta]$ is limit of a sequence of decimal numbers that belong to $[\alpha; \beta]$. Again the fact that T is closed proves that γ belongs to T. To sum up: $T = [\alpha; \beta]$. ∎

4
Continuous functions

4.1 Pointwise continuity

4.1.1 The concept of function

A procedure f that, given x as input, calculates y as output, is called an assignment that is *defined at* x and that appoints the *value* y to the *argument* x with the notation
$$f(x) = y.$$
We say that the assignment f is *defined on* a set U, if
1. f is defined at *all* elements of U, and
2. *only* elements of U are allowed to be arguments of f.

We write $f : U \to V$ if and only if f is defined on U and all values of f are elements of V.

An assignment f is called a *function* if and only if it is *extensional*. By this we mean the following: if f assigns to the argument x' the value $f(x') = y'$, if f assigns to the argument x'' the value $f(x') = y''$, and if the two values are apart, $y' \neq y''$, then the two arguments also are apart: $x' \neq x''$. This condition of course has as its consequence that the relation $x' = x'$, i.e. the equality of arguments at which the assignment is defined, causes the equality of the corresponding values $f(x') = f(x'')$.

Let for instance n denote a positive integer. The procedure that, given the real number α as input, calculates $[\alpha]_n$ as output, is an assignment that is defined on the continuum, but it is *not* a function.

The assignment that appoints to each argument x one and the same value y_0 is of course a function that can be defined on every set. It is called a *constant* function which – "par abus de langage" – can be identified with its only value y_0. The assignment that appoints to each argument x just this argument x is of course a function that can be defined on every set. It is called the *identical* function. Suppose that f and g denote two functions such that the function f is defined at each value of the function g. Then the *concatenation* $f \circ g$ of these two functions is a procedure which is expressed in the formula $f \circ g(x) = f(g(x))$. This assignment $f \circ g$ is a function that is defined at x if and only if g is defined at x. (The notation $f(g)$ instead of $f \circ g$ would be the more suggestive one. But we here retain the usual one.)

Let X and Y denote two sequences

$$X = (x_1, x_2, \ldots, x_n, \ldots), \qquad Y = (y_1, y_2, \ldots, y_n, \ldots).$$

The assignment f that, given $x = x_n$ with a positive integer n as input, calculates $y = y_n$ as output, is an assignment which can be defined on the sequence X. This assignment is a function, and will be called a *function defined as a pair of sequences*, if the sequence X is *entirely discrete*. This means that for all positive integers n, m the inequality $n \neq m$ implies the apartness $x_n \neq x_m$.

This example conveys all possibilities of a function that is defined on a metric space S if this metric space is *enumerable*, i.e. if there exists an entirely discrete sequence of points belonging to S such that each point of S is an element of this sequence. For instance all functions defined on the space of all decimal numbers can be interpreted as functions defined as pairs of sequences.

The "ruler-scale-function" f is defined on the space of all decimal numbers as follows: Any decimal number $a = z + 0.z_1 z_2 \ldots z_n$ is assigned either to

$$f(a) = 10^0 = 1$$

if all digits $z_1, z_2, \ldots, z_n$ are zero, i.e. if a is an integer, or to

$$f(a) = 10^{-k}$$

if $z_k \neq 0$ but $z_j = 0$ for all integers j fulfilling $k < j \leq n$.

Let f and f^* denote two functions, f being defined on U, f^* being defined on U^*. The function f is called *embedded* into the function f^* and f^* is called an *enlargement* of f if and only if
1. U is a subset of U^*, and
2. for each $x \in U$ the equality $f(x) = f^*(x)$ holds.

Two functions f' and f^* are called *equal* if and only if each of them is embedded in the other.

4.1.2 The continuity of a function at a point

Let S and T denote two complete metric spaces. Suppose that the function f is defined on a subset U of S and has values in T. Let further ξ denote a point that belongs to the cover of U. The function f is called *continuous at the point ξ* if and only if it is possible to construct for each real number $\varepsilon > 0$ a real number $\delta > 0$ such that for all points $x', x'' \in U$ fulfilling the inequalities

$$\|x' - \xi\| < \delta \quad \text{and} \quad \|x'' - \xi\| < \delta$$

the inequality

$$\|f(x') - f(x'')\| < \varepsilon$$

holds.

It is clear from the context that in the formulas $\|x' - \xi\| < \delta$ and $\|x'' - \xi\| < \delta$ the metric is the distance defined on S whereas in $\|f(x') - f(x'')\| < \varepsilon$ the metric is the distance defined on T. It is unnecessary to denote these two metrics in different notations.

The clue of this definition of course is that ξ needs not to belong to U, but only to the cover of U. The advantage of defining "continuous" in this special way will soon become clear.

Suppose that the function $f : U \to T$ is continuous at the point ξ which lies in the cover of U. Then there exists a uniquely defined point $\eta \in T$ with the following property: For each sequence X that consists of points $x_1, x_2, \ldots, x_n, \ldots \in U$ and converges to ξ, $\lim X = \xi$, the sequence $Y = f(X)$ that consists of the corresponding values $y_1 = f(x_1)$, $y_2 = f(x_2)$, ..., $y_n = f(x_n)$, ... also is convergent, and we have $\lim Y = \eta$.

Proof. Suppose $X = (x_1, x_2, \ldots, x_n, \ldots)$ consists of points in U and converges to ξ, and let $\varepsilon > 0$ denote an arbitrary real number. The continuity of f at the point ξ allows us to construct a corresponding real number $\delta > 0$. As $\lim X = \xi$, it is further possible to construct a positive integer j such that the inequalities $n \geq j$ and $m \geq j$ imply

$$\|x_n - \xi\| < \delta \quad \text{and} \quad \|x_m - \xi\| < \delta$$

with the consequence

$$\|f(x_n) - f(x_m)\| < \varepsilon .$$

This proves, by the CAUCHY-criterion, that $Y = f(X)$ is a converging sequence, i.e. the limit $\eta = \lim Y$ exists, and is of course a point of the complete metric space T.

We finally show that this point η only depends on the function f and the point ξ, but not on the choice of the sequence X: Suppose we have another sequence X' consisting of points in U which converges to ξ. Then by the same argument that proved the existence of $\lim f(X)$, also $\lim f(X')$ must exist. And the mixed sequence $X^* = X \sqcup X'$ also consists of points in U and converges to ξ. Therefore the sequence $Y^* = f(X^*)$ is convergent and has both, the sequences $Y = f(X)$ and $Y' = f(X')$, as subsequences. This leads to

$$\lim f(X) = \lim f(X^*) = \lim f(X') ,$$

as asserted. ∎

Suppose the point ξ at which the function f is continuous, belongs to U. Then the point η which is constructed by the proposition above, must coincide with $f(\xi)$: The constant sequence X consisting of ξ alone leads to the constant sequence $Y = f(X)$ consisting of $f(\xi)$ alone. We therefore agree to denote the point η which is constructed by the proposition above, even in the general case in which ξ needs only to belong to the cover of U, by $\eta = f(\xi)$.

The constant function (defined on a metric space), for example, is continuous at each point.

The ruler-scale-function, for example, is continuous at every real number ξ that is apart from each decimal number. The reason is as follows: If ξ denotes a real number that is apart from each decimal number, and $\varepsilon > 0$ denotes any real number, it is possible to construct a positive integer j such that $10^{-j} \leq \varepsilon$. The set V of all decimal numbers with exactly j decimal places is a located and closed set in the continuum. By the positive-distance-theorem, it is possible to construct the real number $\delta = \inf |V - \xi|$, and we have $\delta > 0$. For any decimal number x, the inequality $|x - \xi| < \delta$ implies that x cannot be a decimal number with exactly j decimal places, and by definition of the ruler-scale-function $f(x) \leq 10^{-j-1}$. This proves that for all decimal numbers x', x'' the inequalities

$$|x' - \xi| < \delta \quad \text{and} \quad |x'' - \xi| < \delta$$

imply

$$|f(x') - f(x'')| \leq 2 \times 10^{-j-1} < 10^{-j} \leq \varepsilon .$$

This is an instructive example of a function that is *not* continuous at any point where it is defined, but *is* continuous at all limit points that are apart from each point where this function is defined.

4.1.3 Three properties of continuity

Suppose that the function $f : U \to T$ is continuous at the point ξ. Then for each real number $\varepsilon > 0$ it is possible to construct a real number $\delta > 0$ such that for all $x \in U$ the inequality $\|x - \xi\| < \delta$ implies the inequality $\|f(x) - f(\xi)\| < \varepsilon$.

Proof. Let $\varepsilon > 0$ denote an arbitrary real number, and construct a decimal number e such that $\varepsilon > e > 0$. As f is continuous at the point ξ, it is possible to construct a real number $\delta > 0$ such that for all points $x, x' \in U$ the two inequalities

$$\|x - \xi\| < \delta \quad \text{and} \quad \|x' - \xi\| < \delta$$

imply

$$\|f(x) - f(x')\| < e .$$

Suppose that $X = (x_1, x_2, \ldots, x_n, \ldots)$ consists of points in U and converges to ξ. Then there exists a positive integer j such that $n \geq j$ implies that $\|x_n - \xi\| < \delta$.

We therefore have for all integers $n \geq j$ the inequality $\|f(x) - f(x_n)\| < e$ and the sequence consisting of the points $f(x_1), f(x_2), \ldots, f(x_n), \ldots$ converges to $f(\xi)$. The estimation of the limit proves

$$\|f(x) - f(\xi)\| \leq e < \varepsilon .$$

∎

Suppose that f and g denote two functions, defined on metric spaces V and U respectively, such that the function f is defined at each value of the function g. If g is continuous at ξ and if f is continuous at $g(\xi)$, then the concatenation $f \circ g$ is continuous at ξ.

Proof. It is possible to construct for each real number $\varepsilon > 0$ a real number $\rho > 0$ such that for all points $y', y'' \in V$ the inequalities

$$\|y' - g(\xi)\| < \rho \quad \text{and} \quad \|y'' - g(\xi)\| < \rho$$

imply

$$\|f(y') - f(y'')\| < \varepsilon .$$

And it is possible to construct a real number $\delta > 0$ such that for all points x', $x'' \in U$ fulfilling the inequalities

$$\|x' - \xi\| < \delta \quad \text{and} \quad \|x'' - \xi\| < \delta$$

the inequalities

$$\|g(x') - g(\xi)\| < \rho \quad \text{and} \quad \|g(x'') - g(\xi)\| < \rho$$

hold. Therefore the inequalities

$$\|x' - \xi\| < \delta \quad \text{and} \quad \|x'' - \xi\| < \delta$$

imply

$$\|f(g(x')) - f(g(x''))\| < \varepsilon .$$

∎

Suppose that the function $f : U \to T$ and the point $\xi \in U$ fulfill the following condition: For each real number $\varepsilon > 0$ it is possible to construct a real number $\delta > 0$ such that for all $x \in U$ the inequality $\|x - \xi\| < \delta$ implies the inequality $\|f(x) - f(\xi)\| < \varepsilon$. Then f is continuous at the point ξ.

Proof. Let $\varepsilon > 0$ denote an arbitrary real number, and construct a decimal number e such that $\varepsilon > e > 0$. We know that it is possible to construct a real number $\delta > 0$ such that for all points $x \in U$ the inequality $\|x - \xi\| < \delta$ implies that $\|f(x) - f(\xi)\| < e/2$. All the more for all points $x', x'' \in U$ the inequalities

$$\|x' - \xi\| < \delta \quad \text{and} \quad \|x'' - \xi\| < \delta$$

imply
$$\|f(x') - f(\xi)\| < \frac{e}{2} \quad \text{and} \quad \|f(x'') - f(\xi)\| < \frac{e}{2}$$
with the consequence
$$\|f(x') - f(x'')\| \le e < \varepsilon.$$

■

The identical function (defined on a metric space) is continuous at each point.

Proof. Referring to the proposition above, it is possible, after the submission of an arbitrary real number $\varepsilon > 0$, to define $\delta = \varepsilon$. ■

The function $f : U \to T$ is continuous at the point ξ of the cover of U if and only if it is possible to construct for each real number $\varepsilon > 0$ a real number $\delta > 0$ with the following property: If ξ' and ξ'' denote arbitrary points of the cover of U which either belong to U itself or at which the function f is continuous, and if the two inequalities
$$\|\xi' - \xi\| < \delta \quad \text{and} \quad \|\xi'' - \xi\| < \delta$$
hold, then we have the inequality
$$\|f(\xi') - f(\xi'')\| < \varepsilon.$$

Proof. It is obvious that the condition of the proposition implies the continuity of f at the point ξ.

Assume on the other hand that f is continuous at the point ξ, and let $\varepsilon > 0$ denote an arbitrary real number. We now construct a decimal number e such that $\varepsilon > e > 0$. We construct further a decimal number $d > 0$ with the property that for all points $x', x'' \in U$ the two inequalities
$$\|x' - \xi\| < d \quad \text{and} \quad \|x'' - \xi\| < d$$
imply the inequality
$$\|f(x') - f(x'')\| < \frac{e}{2}.$$

We finally define $\delta = d/2$.

Suppose ξ', ξ'' denote points of the cover of U at which f is continuous. This prerequisite allows us to construct two positive decimal numbers d', d'' which, without loss of generality, fulfill the inequalities
$$d' \le \frac{d}{4} \quad \text{and} \quad d'' \le \frac{d}{4},$$
such that for all points $x', x'' \in U$ the inequalities
$$\|x' - \xi'\| < d' \quad \text{resp.} \quad \|x'' - \xi''\| < d''$$

imply the inequalities

$$\|f(x') - f(\xi')\| < \frac{e}{4} \quad \text{resp.} \quad \|f(x'') - f(\xi'')\| < \frac{e}{4}.$$

(The *existence* of points x', x'' of that kind is guaranteed by the fact that ξ', ξ'' belong to the cover of U.) If we assume that the inequalities

$$\|\xi' - \xi\| < \delta \quad \text{and} \quad \|\xi'' - \xi\| < \delta$$

hold, we can derive for all points x', $x'' \in U$ fulfilling $\|x' - \xi'\| < d'$ and $\|x'' - \xi''\| < d''$ the two inequalities $\|x' - \xi\| < d$ and $\|x'' - \xi\| < d$. From this we conclude the three inequalities

$$\|f(x') - f(\xi')\| < \frac{e}{4} \quad \text{and} \quad \|f(x'') - f(\xi'')\| < \frac{e}{4}$$

and also

$$\|f(x') - f(x'')\| < \frac{e}{2}.$$

Thus we have, by the triangle inequality,

$$\|f(\xi') - f(\xi'')\| \leq \frac{e}{4} + \frac{e}{2} + \frac{e}{4} = e < \varepsilon.$$

Suppose ξ' denotes a point of the cover of U at which f is continuous, and $\xi'' = x''$ belongs to U itself. This prerequisite allows us to construct a positive decimal numbers d' which, without loss of generality, fulfills the inequality $d' \leq d/2$, such that for all points $x' \in U$ the inequality $\|x' - \xi'\| < d'$ implies the inequality

$$\|f(x') - f(\xi')\| < \frac{e}{2}.$$

(The *existence* of a point x' of that kind is guaranteed by the fact that ξ' belongs to the cover of U.) If we assume that the inequalities

$$\|\xi' - \xi\| < \delta \quad \text{and} \quad \|\xi'' - \xi\| < \delta$$

hold, we can derive for all points x' of U fulfilling $\|x' - \xi'\| < d'$ the inequality $\|x' - \xi\| < d$, and we also have

$$\|x'' - \xi\| = \|\xi'' - \xi\| < \delta < d.$$

From this we conclude the two inequalities

$$\|f(x') - f(\xi')\| < \frac{e}{2}$$

and

$$\|f(x') - f(\xi'')\| = \|f(x') - f(x'')\| < \frac{e}{2}.$$

Thus we have, by the triangle inequality,

$$\|f(\xi') - f(\xi'')\| \leq \frac{e}{2} + \frac{e}{2} = e < \varepsilon.$$

Suppose finally that $\xi' = x'$, $\xi'' = x''$ denote two points of U and that the two inequalities

$$\|\xi' - \xi\| < \delta \quad \text{and} \quad \|\xi'' - \xi\| < \delta$$

hold. We then have all the more

$$\|x' - \xi\| = \|\xi' - \xi\| < d \quad \text{and} \quad \|x'' - \xi\| = \|\xi'' - \xi\| < d$$

with the consequence

$$\|f(\xi') - f(\xi'')\| = \|f(x') - f(x'')\| \leq \frac{e}{2} < \varepsilon.$$

∎

The last proposition allows us the following construction for a function $f : U \to T$ that is continuous at points ξ of the cover of U: We define U' to be the set of all points at which f is continuous. Then we construct the union $U^* = U \cup U'$, and define the function $f^* : U^* \to T$ to be – according to our agreement about the notation – the assignment which proceeds by setting $f^*(\xi) = f(\xi)$ for all points ξ of U and all points ξ at which f is continuous. f^* is an enlargement of f to U^*. "Par abus de langage" the notation f^* is simplified by dropping the asterisk * and using the original notation f. In any case, the continuity of the enlarged function is inherited from the continuity of the original function.

Suppose for example that U is an enumerable metric space which is dense in the complete metric space S. Suppose further that $f : U \to T$ is a function defined as a pair of sequences, and that f is continuous at each point of S. Then f can be enlarged as continuous function defined on the whole space S, $f : S \to T$. This example is paradigmatic for the *construction* of functions which are defined on *not* enumerable sets.

The ruler-scale-function can be defined on the set of all decimal numbers and of all real numbers ξ that are apart from each decimal number (with the value $f(\xi) = 0$ at these real numbers ξ). But it would be a fallacy to assume that, by doing this, the ruler-scale-function would be defined on the whole continuum: there is no hope to be able to prove that any given real number were either a decimal number or apart from each decimal number.

4.1.4 *Continuity at inner points*

Theorem of WEYL **and** BROUWER. *Let S and T denote complete metric spaces. Let $f : U \to T$ denote a function that is defined on a subset U of S, and suppose that ξ is an inner point of U. Then the function f is continuous at ξ.*

Proof. We assume that S is a complete metric space with (Σ, E) as adequate pair of the two sequences

$$\Sigma = (S_1, S_2, \ldots, S_n, \ldots) \quad \text{and} \quad E = (e_1, e_2, \ldots, e_n, \ldots),$$

and that T is a complete metric space with $(\bar{\Sigma}, \bar{E})$ as adequate pair of the two sequences

$$\bar{\Sigma} = (T_1, T_2, \ldots, T_n, \ldots) \quad \text{and} \quad \bar{E} = (\bar{e}_1, \bar{e}_2, \ldots, \bar{e}_n, \ldots).$$

ξ is presupposed to be an inner point of U. Therefore there exists a positive decimal number d such that the neighborhood $U_d(\xi)$ is a subset of U.

The assignment f that is presupposed in the theorem must be a precisely defined *procedure*. If this procedure assigns the value η to the argument ξ, it must be able to start after the announcement of *finitely* many of the points $[\xi]_1, [\xi]_2, \ldots, [\xi]_n, \ldots$ that define the argument ξ – the notion one would need infinitely many data to start the calculation of η is nonsense. But it would equally be nonsense to expect that the procedure would produce *all* the points $[\eta]_1, [\eta]_2, \ldots, [\eta]_n, \ldots$ at once. What we must expect, however, is that – if ξ is submitted – the procedure calculates for each positive integer j the point $[\eta]_j$. This is the essential condition to ascertain the point η from the sequence of the points $[\eta]_1, [\eta]_2, \ldots, [\eta]_n, \ldots$.

So we have to analyze the phrase: "If j denotes an arbitrary positive integer, and if ξ is submitted, then the procedure calculates $[\eta]_j$." Whereby the assignment refers explicitly to the submission of ξ and *not* to the submission of the special sequence of points $[\xi]_1, [\xi]_2, \ldots, [\xi]_n, \ldots$.

Therefore the word "if ξ is submitted" in the phrase above means the following: *There exists a positive integer k such that, for any points $x_1 \in S_1$ with $\|x_1 - \xi\| \leq e_1$, $x_2 \in S_2$ with $\|x_2 - \xi\| \leq e_2$, ..., $x_k \in S_k$ with $\|x_k - \xi\| \leq e_k$, the assignment f allows the calculation of $[\eta]_j$.*

The justification for this is: we have for all positive integers $n \leq k$ and $m \leq k$, by the triangle inequality,

$$\|x_n - x_m\| \leq e_n + e_m \,\text{,}$$

and we have further for all positive integers $n \leq k$ and $m > k$ from

$$\|[\xi]_m - \xi\| \leq e_m$$

and, again by the triangle inequality,

$$\|x_n - [\xi]_m\| \leq e_n + e_m.$$

This proves that ξ', defined by $[\xi']_1 = x_1$, $[\xi']_2 = x_2$, ..., $[\xi']_k = x_k$, and by $[\xi']_m = [\xi]_m$ for $m > k$, is a point of S fulfilling $\xi' = \xi$. On the other hand: it is possible to identify for each point $\xi' \in S$ with $\xi' = \xi$ the k points $[\xi']_1, [\xi']_2, \ldots, [\xi']_k$ with the corresponding k points $x_1, x_2, \ldots, x_k$ from above because of the identity $\xi' = \xi$ and the approximation lemma

$$\|[\xi']_1 - \xi\| \leq e_1, \quad \|[\xi']_2 - \xi\| \leq e_2, \quad \ldots, \quad \|[\xi']_k - \xi\| \leq e_k.$$

104   4. Continuous functions

And the fact that f refers to the submission of ξ, and *not* to the submission of the special sequence of points $[\xi]_1, [\xi]_2, \ldots, [\xi]_n, \ldots$, has as its consequence that the assignment goes along the same schematic procedure, independent of the fact that the argument ξ or the argument $\xi' = \xi$ is given.

It is of course possible that the procedure assigns to the submitted ξ the point $[\eta]_j$ and assigns to the submitted ξ' with $\xi' = \xi$ the point $[\eta']_j$, but the extensionality of f requests $\eta = \eta'$. This, together with

$$\|[\eta]_j - \eta\| \leq \bar{e}_j \quad \text{and} \quad \|[\eta']_j - \eta'\| \leq \bar{e}_j,$$

leads to the inequality

$$\|[\eta]_j - [\eta']_j\| \leq 2\bar{e}_j.$$

Now suppose that $\varepsilon > 0$ denotes an arbitrary real number. There exists a positive integer j such that $4\bar{e}_j < \varepsilon$. Given this j, there exists a positive integer k such that for any points $x_1 \in S_1$ with $\|x_1 - \xi\| \leq e_1$, $x_2 \in S_2$ with $\|x_2 - \xi\| \leq e_1, \ldots,$ $x_k \in S_k$ with $\|x_k - \xi\| \leq e_1$, the assignment f allows the calculation of $[\eta]_j$. We finally define the positive decimal number δ by

$$\delta = \min\left(d, \frac{e_1 - d_1}{2}, \ldots, \frac{e_k - d_k}{2}\right).$$

Let $x \in S$ denote an arbitrary point such that

$$\|x - \xi\| < \delta.$$

The inequality $\delta \leq d$ implies that $x \in U$. We further define x^* and ξ^* to be the points $x^* = x$ and $\xi^* = \xi$ which can be calculated by the rounding lemma. We finally define for all positive integers $n \leq k$ the points $[\xi']_n = [x^*]_n$, and for all positive integers $n > k$ the points $[\xi']_n = [\xi^*]_n$. The overlapping lemma guarantees that $\xi' \in S$ coincides with ξ. The procedure of the function f allows us, after the submission of $[x^*]_1, [x^*]_2, \ldots, [x^*]_k$, i.e. of $[\xi']_1, [\xi']_2, \ldots, [\xi']_k$, the calculation of $[\eta']_j$. And the equality $x^* = x$ implies that

$$\|[\eta']_j - f(x)\| \leq \bar{e}_j.$$

This together with the two inequalities

$$\|[\eta]_j - [\eta']_j\| \leq 2\bar{e}_j \quad \text{and} \quad \|[\eta]_j - \eta\| = \|[\eta]_j - f(\xi)\| \leq \bar{e}_j$$

from above lead, by the triangle inequality, to $\|f(\xi) - f(x)\| \leq 4\bar{e}_j$, thus to

$$\|f(x) - f(\xi)\| < \varepsilon.$$

∎

This theorem formulates LEIBNIZ's hypothesis "natura non facit saltus" within the context of mathematics. Its proof of course is based on the rejection of the hypothesis of DEDEKIND and CANTOR. For instance the "signum-function" f defined by the assignment

$$f(x) = \begin{cases} -1 & \text{if } x < 0 \\ 0 & \text{if } x = 0 \\ 1 & \text{if } x > 0 \end{cases}$$

is not continuous at the point 0 which of course does not contradict the theorem of BROUWER and WEYL, because it is, within intuitionism, wrong that the whole continuum could be separated into the set of the real numbers $x < 0$, the set consisting only of 0 and the set of the real numbers $x > 0$. Here ARISTOTLE's ancient dictum that "the whole is more than the sum of its parts" is reflected mathematically.

4.2 Uniform continuity

4.2.1 Pointwise and uniform continuity

Let S and T be complete metric spaces. Let $f : U \to T$ denote a function that is defined on a subset U of S.

The function f is called *(pointwise) continuous on U* if it is continuous at all points of U, i.e. if it is possible to construct to each real number $\varepsilon > 0$ and to each point $\xi \in U$ a real number $\delta > 0$ such that, for all points $x \in U$, the inequality $\|x - \xi\| < \delta$ implies the inequality $\|f(x) - f(\xi)\| < \varepsilon$.

The function f is called *uniformly* continuous on U if it is possible to construct to each real number $\varepsilon > 0$ a real number $\delta > 0$ such that, for all points $\xi \in U$ and for all points $x \in U$, the inequality $\|x - \xi\| < \delta$ implies the inequality $\|f(x) - f(\xi)\| < \varepsilon$.

Suppose that $f : U \to T$ is uniformly continuous on U. Then f is continuous at all points ξ belonging to the cover of U, i.e. f is continuous on the cover of U.

Proof. Let $\varepsilon > 0$ denote an arbitrary real number. Then it is possible to construct a real number $\delta > 0$ such that, for all points $x', x'' \in U$, the inequality

$$\|x' - x''\| < \delta$$

implies the inequality

$$\|f(x') - f(x'')\| < \varepsilon .$$

Let d denote a positive decimal number fulfilling $2d \leq \delta$. The two inequalities

$$\|x' - \xi\| < d \quad \text{and} \quad \|x'' - \xi\| < d$$

lead to $\|x' - x''\| < \delta$ and thus to

$$\|f(x') - f(x'')\| < \varepsilon.$$

This, by definition, proves the continuity of f at ξ. ∎

Suppose that $f : U \to T$ is uniformly continuous on U. Then the enlargement of f to the cover of U is uniformly continuous on the cover of U.

Proof. Let $\varepsilon > 0$ denote an arbitrary real number. We construct a positive decimal number $e < \varepsilon$, and we construct a positive decimal number d such that, for all points $x', x'' \in U$, the inequality $\|x' - x''\| < d$ implies that

$$\|f(x') - f(x'')\| < \frac{e}{2}.$$

We now set $\delta = d/2$ and choose two arbitrary points ξ', ξ'' from the cover of U fulfilling $\|\xi' - \xi''\| < \delta$.
Since f is continuous at the point ξ', there exists a positive decimal number $d' \leq d/4$ such that, for all points $x' \in U$ fulfilling $\|x' - \xi'\| < d'$ (and points with this property do exist), the inequality

$$\|f(x') - f(\xi')\| < \frac{e}{4}$$

holds. Since f is continuous at the point ξ'', there exists a positive decimal number $d'' \leq d/4$ such that, for all points $x'' \in U$ fulfilling $\|x'' - \xi''\| < d''$ (and, as before, points with this property do exist), the inequality

$$\|f(x'') - f(\xi'')\| < \frac{e}{4}$$

holds.
The two prerequisites

$$\|x' - \xi'\| < \frac{d}{4} \quad \text{and} \quad \|x'' - \xi''\| < \frac{d}{4}$$

together with $\|\xi' - \xi''\| < d/2$ imply, by the triangle inequality, $\|x' - x''\| < d$, and therefore

$$\|f(x') - f(x'')\| < \frac{\varepsilon}{2}.$$

This combined with the inequalities

$$\|f(x') - f(\xi')\| < \frac{e}{4} \quad \text{and} \quad \|f(x'') - f(\xi'')\| < \frac{e}{4},$$

again by the triangle inequality, has $\|f(\xi') - f(\xi'')\| < \varepsilon$ as a consequence. ∎

4.2.2 Uniform continuity and totally boundness

Suppose that $f : U \to T$ is uniformly continuous on a totally bounded set U. Then the set $V = f(U)$ consisting of all values $y = f(x)$ with arguments $x \in U$ is totally bounded.

Proof. Let $\varepsilon > 0$ denote an arbitrary real number. Then it is possible to construct a real number $\delta > 0$ such that, for all points $x', x'' \in U$, the inequality

$$\|x' - x''\| < \delta$$

implies the inequality

$$\|f(x') - f(x'')\| < \varepsilon.$$

Let $(x_1, x_2, \ldots, x_n)$ denote a finite δ-net of U, and define for all positive integers $j \leq n$ the points y_j to be $y_j = f(x_j)$. If y is an arbitrary point of $V = f(U)$, then there exists a point $x \in U$ such that $y = f(x)$. It is possible to pick up a positive integer $j \leq n$ such that $\|x_j - x\| < \delta$. This implies that $\|f(x_j) - f(x)\| < \varepsilon$, i.e. $\|y_j - y\| < \varepsilon$. Thus the finite sequence $(y_1, y_2, \ldots, y_n)$ is seen to be a finite ε-net of V. ∎

A function is called a *real function* if and only if all its values are real numbers.

Theorem about the supremum and the infimum of real functions. *Suppose that the real function f is defined and uniformly continuous on a totally bounded set U. Then there exist two real numbers $\sigma = \sup f(U)$ and $\rho = \inf f(U)$ with the following properties:*
1. σ is an upper bound and ρ is a lower bound of $f(U)$, i.e. we have the inequalities $\rho \leq f(x) \leq \sigma$ for all $x \in U$,
2. σ is the smallest upper bound, and ρ is the greatest lower bound of $f(U)$, i.e. for all real numbers $\alpha < \sigma$ (resp. $\alpha > \rho$) it is possible to find a point $x_0 \in U$ such that $f(x_0) > \alpha$ (resp. $f(x_0) < \alpha$).

There is no hope that we could prove the existence of points ξ' or ξ'' (that belong to U or to the cover of U) such that $f(\xi') = \sigma$ or $f(\xi'') = \rho$, although we can always find, given an arbitrary small real $\varepsilon > 0$, a point $x' \in U$ such that

$$\sigma - \varepsilon < f(x') \leq \sigma$$

and a point $x'' \in U$ such that

$$\rho \leq f(x'') < \rho + \varepsilon.$$

4.2.3 Uniform continuity and connectedness

Suppose that $f : U \to T$ is uniformly continuous, and the set U connects the two points ξ' and ξ'' of the cover of U. Then the set $V = f(U)$ connects the two values $\eta' = f(\xi')$ and $\eta'' = f(\xi'')$.

Proof. Let $\varepsilon > 0$ denote an arbitrary real number. Then it is possible to construct a real number $\delta > 0$ such that, for all points $x', x'' \in U$, the inequality

$$\|x' - x''\| < \delta$$

implies the inequality

$$\|f(x') - f(x'')\| < \varepsilon .$$

There exists a δ-connection of points $x_0, x_1, x_2, \ldots, x_j \in U$ such that the inequalities

$$\|\xi' - x_0\| < \delta \quad \text{and} \quad \|x_j - \xi''\| < \delta$$

hold, and, by definition, $\|x_{n-1} - x_n\| < \delta$ also holds for each positive integer $n \leq j$. We now define for each nonnegative integer $n \leq j$ the points $y_n = f(x_n)$ and derive on the one hand, for each positive integer $n \leq j$, the inequality

$$\|y_{n-1} - y_n\| = \|f(x_{n-1}) - f(x_n)\| < \varepsilon ,$$

and on the other hand

$$\|\eta' - y_0\| = \|f(\xi') - f(x_0)\| < \varepsilon$$

and

$$\|y_n - \eta''\| = \|f(x_n) - f(\xi'')\| < \varepsilon .$$

Therefore the finite sequence consisting of $y_0, y_1, y_2, \ldots, y_j$ is an ε-connection of V that connects η' with η''. ∎

Suppose that $f : U \to T$ is uniformly continuous on a connected set U. Then the set $V = f(U)$ consisting of all values $y = f(x)$ with arguments $x \in U$ is connected.

Suppose that the real function f is defined and uniformly continuous on a totally bounded and connected set U. Then the cover of $V = f(U)$ coincides with the compact interval $[\rho; \sigma]$ with the borders $\rho = \inf f(U)$, $\sigma = \sup f(U)$.

Proof. V is a totally bounded and connected set in the continuum This implies that the cover of V is a compact and connected set in the continuum and therefore a compact interval with the borders $\rho = \inf V$, $\sigma = \sup V$. ∎

Theorem of BOLZANO. *Suppose that the real function f is defined and uniformly continuous on a totally bounded and connected set U, and define*

$$\rho = \inf f(U) , \quad \sigma = \sup f(U) .$$

Then, for any real $\varepsilon > 0$ and any real number $\eta \in [\rho; \sigma]$, it is possible to construct a point $x_0 \in U$ such that $|f(x_0) - \eta| < \varepsilon$.

Nevertheless there is, without any further prerequisites, no hope that it would be possible to prove the existence of a point ξ belonging to the cover of U such that $f(\xi) = \eta$.

4.2.4 Uniform continuity on compact spaces

Theorem of BROUWER. *Suppose that S is a compact metric space and T is a complete metric space. Then each function $f : S \to T$ is uniformly continuous on S.*

Proof. We assume (Σ, E) to be an adequate pair of the two sequences

$$\Sigma = (S_1, S_2, \ldots, S_n, \ldots) \quad \text{and} \quad E = (e_1, e_2, \ldots, e_n, \ldots)$$

for the compact metric space S such that, for all positive integers n, the sets S_n are finite.

As each point of S is an inner point of S, the function f is pointwise continuous on S. Suppose that $\varepsilon > 0$ denotes an arbitrary real number. There exists a positive decimal number $e < \varepsilon$, and to each point $\xi \in S$ it is possible to construct a positive decimal number d such that, for all points $x \in S$, the inequality $\|x - \xi\| < d$ implies the inequality $\|f(x) - f(\xi)\| < e/2$. Let k be a positive integer such that $e_k \leq d$. We then have, for all positive integers $n \geq k$, by the approximation lemma

$$\|[\xi]_n - \xi\| \leq e_n \leq e_k \leq d,$$

and therefore

$$\|f([\xi]_n) - f(\xi)\| < \frac{e}{2}.$$

To put it into one sentence: *To each point $\xi \in S$ it is possible to construct a positive integer k such that, for all positive integers $n \geq k$, the inequality $\|f([\xi]_n) - f(\xi)\| < e/2$ holds.*

We now define W to be a subset of the union of the sets $S_1, S_2, \ldots, S_n, \ldots$ with the following property: if ξ denotes an arbitrary point of S and n denotes a positive integer, the fact that $[\xi]_n = x$ belongs to W follows from $\|f(x) - f(\xi)\| < e/2$ and vice-versa. The conclusion from above implies that W is a *bar*. We derive from the bar-theorem the existence of a positive integer m such that, for all points $\xi \in S$ and all positive integers $n \geq m$, the inequality $\|f([\xi]_n) - f(\xi)\| < e/2$ holds. Finally the positive decimal number δ is defined to be

$$\delta = \min\left(\frac{e_1 - d_1}{2}, \ldots, \frac{e_m - d_m}{2}\right).$$

Let ξ and x denote two arbitrary points of S with the property $\|\xi - x\| < \delta$. By the rounding lemma it is possible to assign to ξ a rounded approximation sequence ξ^* fulfilling $\xi^* = \xi$, to assign to x a rounded approximation sequence x^* fulfilling $x^* = x$, and, by the overlapping lemma, the definitions $[x']_n = [\xi^*]_n$ for all $n \leq m$ and $[x']_n = [x^*]_n$ for all $n > m$ fix a point x' of S such that $x' = x$. The two inequalities

$$\|f([\xi^*]_m) - f(\xi)\| = \|f([\xi^*]_m) - f(\xi^*)\| < \frac{e}{2}$$

and
$$\|f([\xi^*]_m) - f(x)\| = \|f([x']_m) - f(x')\| < \frac{e}{2}$$
imply, by the triangle inequality, $\|f(\xi) - f(x)\| \le e < \varepsilon$. ∎

4.3 Elementary calculations in the continuum

4.3.1 Continuity of addition and multiplication

The real function f defined on $\mathbb{D}^2$ by the assignment $f(x, y) = x + y$ is continuous on $\mathbb{R}^2$.

Proof. Let (ξ, η) be an arbitrary point of $\mathbb{R}^2$, and let $\varepsilon > 0$ be an arbitrary real number. We construct a positive decimal number $e < \varepsilon$ and define $\delta = e/4$. The inequalities
$$\|(x', y') - (\xi, \eta)\|_\infty < \delta \quad \text{and} \quad \|(x'', y'') - (\xi, \eta)\|_\infty < \delta$$
imply the four inequalities
$$|x' - \xi| < \frac{e}{4} \qquad |x'' - \xi| < \frac{e}{4}$$
$$|y' - \eta| < \frac{e}{4} \qquad |y'' - \eta| < \frac{e}{4}$$
Thus, by the triangle inequality, we have
$$|x' - x''| \le \frac{e}{2} \quad \text{and} \quad |y' - y''| \le \frac{e}{2}$$
with the consequence
$$\begin{aligned}|(x' + y') - (x'' + y'')| &= |(x' - x'') + (y' - y'')| \\ &\le |x' - x''| + |y' - y''| \le e < \varepsilon.\end{aligned}$$
∎

The real function f defined on $\mathbb{D}^2$ by the assignment $f(x, y) = xy$ is continuous on $\mathbb{R}^2$.

Proof. Let (ξ, η) be an arbitrary point of $\mathbb{R}^2$, and let $\varepsilon > 0$ be an arbitrary real number. We choose a positive integer k so big that $|\xi - x| \le 1$ and $|\eta - y| \le 1$ imply $|x| \le 10^k$ and $|y| \le 10^k$, we construct a positive decimal number $e < \varepsilon$, and we define
$$\delta = \min\left(\frac{e}{4} \times 10^{-k}, 1\right).$$
The inequalities
$$\|(x', y') - (\xi, \eta)\|_\infty < \delta \quad \text{and} \quad \|(x'', y'') - (\xi, \eta)\|_\infty < \delta$$

imply the four inequalities

$$|x' - \xi| < \frac{e}{4} \times 10^{-k} \qquad |x'' - \xi| < \frac{e}{4} \times 10^{-k}$$
$$|y' - \eta| < \frac{e}{4} \times 10^{-k} \qquad |y'' - \eta| < \frac{e}{4} \times 10^{-k}$$

Thus, by the triangle inequality, we have

$$|x' - x''| \leq \frac{e}{2} \times 10^{-k} \quad \text{and} \quad |y' - y''| \leq \frac{e}{2} \times 10^{-k}$$

with the consequence

$$\begin{aligned}
|x'y' - x''y''| &= |(x'y' - x''y') + (x''y' - x''y'')| \\
&\leq |y'||x' - x''| + |x''||y' - y''| \\
&\leq 10^k \times \frac{e}{2} \times 10^{-k} + 10^k \times \frac{e}{2} \times 10^{-k} = e < \varepsilon.
\end{aligned}$$

∎

It is possible to assign to each pair of real numbers α, β a sum $\alpha + \beta$ and a product $\alpha\beta$ (which are real numbers) such that
1. these assignments are enlargements of sum and product defined on decimal numbers,
2. these assignments obey for any real numbers α, β, γ the algebraic rules

$$\begin{array}{ll}
\alpha + (\beta + \gamma) = (\alpha + \beta) + \gamma, & \alpha(\beta\gamma) = (\alpha\beta)\gamma, \\
\alpha + \beta = \beta + \alpha, & \alpha\beta = \beta\alpha, \\
\alpha - \beta = \alpha + (-1)\beta, & \alpha(\beta + \gamma) = \alpha\beta + \alpha\gamma, \\
\alpha + 0 = \alpha, & 1\alpha = \alpha,
\end{array}$$

3. for any real numbers α, β, the inequalities $\alpha > 0$, $\beta > 0$ imply the inequalities $\alpha + \beta > 0$, $\alpha\beta > 0$.

Proof. 1: This is a consequence of the continuity of these operations on $\mathbb{R}^2$.
2: This follows from the corresponding rules for decimal numbers and the principle of permanence.
3: The inequalities $\alpha > 0$, $\beta > 0$ allow us to construct two decimal numbers a, b such that $\alpha > a > 0$ and $\beta > b > 0$. By the principle of permanence we have $\alpha + \beta \geq a + b$, $\alpha\beta \geq ab$, and the two inequalities $a + b > 0$, $ab > 0$ immediately imply $\alpha + \beta > 0$, $\alpha\beta > 0$. ∎

4.3.2 Continuity of the absolute value

The real function f defined on $\mathbb{D}$ by the assignment $f(x) = |x|$ is continuous on $\mathbb{R}$.

Proof. Let ξ be an arbitrary point of $\mathbb{R}$, and let $\varepsilon > 0$ be an arbitrary real number. We construct a positive decimal number $e < \varepsilon$ and define $\delta = e/2$. The inequalities

$$|x' - \xi| < \frac{e}{2} \qquad |x'' - \xi| < \frac{e}{2}$$

imply, by the triangle inequality, $|x' - x''| \leq e$ with the consequence

$$||x'| - |x''|| \leq |x' - x''| \leq e < \varepsilon,$$

as asserted. ∎

It is possible to assign to each pair of real numbers α, β a maximum $\max(\alpha, \beta)$ and a minimum $\min(\alpha, \beta)$ (which are real numbers) such that
1. these assignments are enlargements of maximum and minimum defined on decimal numbers,
2. these assignments obey for any real numbers α, β the rules

$$\begin{aligned}
\alpha \leq \max(\alpha, \beta), &\qquad \min(\alpha, \beta) \leq \alpha, \\
\max(\alpha, \beta) = \max(\beta, \alpha), &\qquad \min(\alpha, \beta) = \min(\beta, \alpha), \\
|\alpha| = \max(\alpha, -\alpha), &\qquad -|\alpha| = \min(\alpha, -\alpha),
\end{aligned}$$

3. for any real numbers α, β, γ, the inequality $\alpha \leq \beta$ implies that

$$\max(\alpha, \beta) = \beta \quad \text{and} \quad \min(\alpha, \beta) = \alpha.$$

Furthermore, the inequality $\max(\alpha, \beta) > \gamma$ implies at least one of the relations $\alpha > \gamma$ or $\beta > \gamma$, and the inequality $\min(\alpha, \beta) < \gamma$ implies at least one of the relations $\alpha < \gamma$ or $\beta < \gamma$.

Proof. 1: The two assignments max and min defined on $\mathbb{D}^2$ by

$$\max(x, y) = \frac{1}{2}(x + y + |x - y|), \qquad \min(x, y) = \frac{1}{2}(x + y - |x - y|)$$

are continuous on $\mathbb{R}^2$.
2: This follows from the corresponding rules for decimal numbers and the principle of permanence.
3: The inequality $\alpha \leq \beta$ implies that $|\alpha - \beta| = \beta - \alpha$. Furthermore, the inequality $\max(\alpha, \beta) > \gamma$ allows us to construct a decimal number c such that $\max(\alpha, \beta) > c > \gamma$. By the dichotomy lemma, at least one of the relations $\alpha > \gamma$ or $c > \alpha$ must be true, and, by the same lemma, at least one of the relations $\beta > \gamma$ or $c > \beta$ must be true. But the assumption $c > \alpha$ and $c > \beta$ implies the absurdity $c \geq \max(\alpha, \beta)$, thus at least one of the relations $\alpha > \gamma$ or $\beta > \gamma$ must be true. The argument for $\min(\alpha, \beta) < \gamma$ is, mutatis mutandis, the same. ∎

Let, as a first typical example, α denote a pendulum number such that $|\alpha| < 1$. The function $f : [-2; 2] \to \mathbb{R}$ is defined by the assignment

$$f(x) = \begin{cases} \alpha + 1 + x & \text{if } -2 \leq x \leq -1 \\ \alpha & \text{if } -1 \leq x \leq 1 \\ \alpha - 1 + x & \text{if } 1 \leq x \leq 2 \end{cases}$$

Although 0 is contained in $f([-2; 2]) = [\alpha - 1; \alpha + 1]$, the proof of the existence of a real number $\xi \in [-2; 2]$ such that $f(\xi) = 0$ would imply that one could decide between the possibilities $\alpha \geq 0$ or $\alpha \leq 0$ – an untenable assumption for an arbitrary pendulum number α.

Let, as a second illustrative example, α denote a pendulum number such that $|\alpha| < 1$. The function $f : [-2; 2] \to \mathbb{R}$ is defined by the assignment

$$f(x) = \begin{cases} (1 + \alpha)(2 + x) & \text{if } -2 \leq x \leq -1 \\ (1 + \alpha)(-x) & \text{if } -1 \leq x \leq 0 \\ (1 - \alpha)x & \text{if } 0 \leq x \leq 1 \\ (1 - \alpha)(2 - x) & \text{if } 1 \leq x \leq 2 \end{cases}$$

Although we have $\sup f([-2; 2]) = 1 + |\alpha|$, the proof of the existence of a real number $\xi \in [-2; 2]$ such that $f(\xi) = 1 + |\alpha|$ would imply that one could decide between the possibilities $\alpha \geq 0$ or $\alpha \leq 0$ – an untenable assumption for an arbitrary pendulum number α.

4.3.3 Continuity of division

To each decimal number a and to each decimal number $b \neq 0$ there exists a uniquely determined real number γ such that $b\gamma = a$; this real number is denoted as fraction $\gamma = a/b$.

Proof. Let l denote a positive integer with the following property: either that $p = a \times 10^l$ is an integer and $m = b \times 10^l$ is a positive integer, or that $p = -a \times 10^l$ is an integer and $m = -b \times 10^l$ is a positive integer. The relations $b\gamma = a$ and $m\gamma = p$ of course are equivalent.

We now refer to the fact that, for each integer p and each positive integer m, there exist two (uniquely determined) integers q and r such that

$$p = mq + r \quad \text{and} \quad 0 \leq r < m.$$

We can even construct, for any positive integer n, two (uniquely determined) integers q_n and r_n such that

$$10^n \times p = mq_n + r_n \quad \text{and} \quad 0 \leq r_n < m.$$

We now set $[\gamma]_n = q_n \times 10^{-n}$ and $c_n = r_n \times 10^{-n}$. We have for any positive integers n and k the relations

$$p = [\gamma]_n m + c_n, \qquad 0 \leq c_n < m \times 10^{-n},$$

and

$$p = [\gamma]_{n+k} m + c_{n+k}, \qquad 0 \leq c_{n+k} < m \times 10^{-n-k},$$

with the consequence

$$m([\gamma]_{n+k} - [\gamma]_n) = c_n - c_{n+k}$$

from which the inequality

$$\left|[\gamma]_{n+k} - [\gamma]_n\right| \leq 10^{-n} + 10^{-n-k}$$

follows. This proves that γ defines a real number. From the relations

$$p \leq m[\gamma]_n < p + m \times 10^{-n}$$

which hold for all positive integers n, we conclude $p \leq m\gamma \leq p$ by the permanence principle. This proves $m\gamma = p$, i.e. $b\gamma = a$.

γ is uniquely determined for the following reason: Suppose that $b > 0$ holds. For any γ' that is apart form γ, we derive from $\gamma' < \gamma$ the relation $b\gamma' < a$, and from $\gamma' > \gamma$ the relation $b\gamma' > a$. Suppose that $b < 0$ holds. For any γ' that is apart form γ, we equally derive from $\gamma' < \gamma$ the relation $b\gamma' > a$, and from $\gamma' > \gamma$ the relation $b\gamma' < a$. ∎

The real function f defined on the decimal numbers being different from zero by the assignment $f(x) = 1/x$ is continuous on the set of all real numbers that are apart from zero.

Proof. Let $\xi \neq 0$ be an arbitrary point of $\mathbb{R}$, and let $\varepsilon > 0$ be an arbitrary real number. We construct a positive decimal number a such that $2a < |\xi|$. This guarantees for all decimal numbers x that $|\xi - x| \leq |\xi| - a$ implies that $|x| \geq a$. We further construct a positive decimal number $e < \varepsilon$ and define

$$\delta = \min\left(\frac{a^2 e}{2}, |\xi| - a\right).$$

The inequalities

$$|x' - \xi| < \delta \quad \text{and} \quad |x'' - \xi| < \delta$$

imply, by the triangle inequality,

$$|x' - x''| \leq a^2 e.$$

Furthermore, the relation $\delta \leq |\xi| - a$ implies that $|x'| \geq a$ and $|x''| \geq a$. This has the consequence

$$\left|\frac{1}{x'} - \frac{1}{x''}\right| = \left|\frac{x'' - x'}{x'x''}\right| = \frac{1}{|x'||x''|}|x' - x''|$$
$$\leq \frac{1}{a^2} \cdot a^2 e = e < \varepsilon.$$

∎

To each real number α and to each real number $\beta \neq 0$ there exists a uniquely determined real number γ such that $\beta\gamma = \alpha$; this real number is denoted as fraction $\gamma = \alpha/\beta$ which abbreviates $\gamma = \alpha(1/\beta)$.

4.3.4 Inverse functions

A real function f defined on a set U of the continuum is called
1. *strictly monotone increasing* if and only if, for any real numbers $x', x'' \in U$, the inequality $x' < x''$ implies that $f(x') < f(x'')$, and *strictly monotone decreasing* if and only if, for any real numbers $x', x'' \in U$, the inequality $x' < x''$ implies that $f(x') > f(x'')$,
2. *monotone increasing* if and only if, for any real numbers $x', x'' \in U$, the inequality $x' < x''$ implies that $f(x') \leq f(x'')$, and *monotone decreasing* if and only if, for any real numbers $x', x'' \in U$, the inequality $x' < x''$ implies that $f(x') \geq f(x'')$.

The function is called *strictly monotone* if and only if it is strictly monotone increasing or strictly monotone decreasing. The function is called *monotone* if and only if it is monotone increasing or monotone decreasing.

Intermediate value theorem for strictly monotone functions. *Let α, β denote two real numbers such that $\alpha < \beta$ and assume that the real function f is defined on the compact interval $[\alpha; \beta]$. If f is strictly monotone increasing, then it is possible to construct for any real number $\eta \in [f(\alpha); f(\beta)]$ the uniquely determined real number $\xi \in [\alpha; \beta]$ with the property $f(\xi) = \eta$. If f is strictly monotone decreasing, then it is possible to construct for any real number $\eta \in [f(\beta); f(\alpha)]$ the uniquely determined real number $\xi \in [\alpha; \beta]$ with the property $f(\xi) = \eta$.*

Proof. We formulate the proof for strictly monotone increasing functions; the analogous proof for strictly monotone decreasing functions follows by interchanging the corresponding inequality-signs.
We construct two sequences A, B, the sequence A consisting of real numbers $\alpha_1, \alpha_2, \ldots, \alpha_n, \ldots$ such that

$$\alpha_1 \leq \alpha_2 \leq \ldots \leq \alpha_n \leq \ldots,$$

the sequence B consisting of real numbers $\beta_1, \beta_2, \ldots, \beta_n, \ldots$ such that

$$\beta_1 \geq \beta_2 \geq \ldots \geq \beta_n \geq \ldots,$$

with the following additional properties:
1. For all positive integers n the inequality $\alpha_n < \beta_n$ holds.
2. For all positive integers n at least one of the two inequalities

$$f(\alpha_n) \leq \eta < f(\beta_n) \quad \text{or} \quad f(\alpha_n) < \eta \leq f(\beta_n)$$

holds.
3. If we define $\alpha_0 = \alpha$, $\beta_0 = \beta$, we have for all positive integers n the relation

$$\beta_n - \alpha_n = \frac{2}{3}(\beta_{n-1} - \alpha_{n-1}).$$

4. Continuous functions

Suppose by induction that, for some positive integer n, we have already calculated $\alpha_0, \ldots, \alpha_{n-1}$ and $\beta_0, \ldots, \beta_{n-1}$, and these numbers possess the above mentioned properties. We now define

$$\alpha' = \frac{2\alpha_{n-1} + \beta_{n-1}}{3}, \quad \beta' = \frac{\alpha_{n-1} + 2\beta_{n-1}}{3}.$$

This definition implies that

$$\alpha_{n-1} < \alpha' < \beta' < \beta_{n-1} \quad \text{and} \quad f(\alpha_{n-1}) < f(\alpha') < f(\beta') < f(\beta_{n-1}).$$

By the dichotomy lemma, at least one of the two inequalities $\eta < f(\beta')$ or $f(\alpha') < \eta$ must hold. If we have $\eta < f(\beta')$, we obviously have

$$f(\alpha_{n-1}) \leq \eta < f(\beta_{n-1}).$$

In this case we define $\alpha_n = \alpha_{n-1}, \beta_n = \beta'$, with the consequences

$$\alpha_{n-1} \leq \alpha_n, \quad \beta_{n-1} \geq \beta_n, \quad \alpha_n < \beta_n, \quad f(\alpha_n) \leq \eta < f(\beta_n),$$

and

$$\beta_n - \alpha_n = \frac{2}{3}(\beta_{n-1} - \alpha_{n-1}).$$

If we have $f(\alpha') < \eta$, we obviously have

$$f(\alpha_{n-1}) < \eta \leq f(\beta_{n-1}).$$

In this case we define $\alpha_n = \alpha', \beta_n = \beta_{n-1}$, with the consequences

$$\alpha_{n-1} \leq \alpha_n, \quad \beta_{n-1} \geq \beta_n, \quad \alpha_n < \beta_n, \quad f(\alpha_n) < \eta \leq f(\beta_n),$$

and

$$\beta_n - \alpha_n = \frac{2}{3}(\beta_{n-1} - \alpha_{n-1}).$$

Now we prove the convergence of A, of B, even of the mixed sequence $A \sqcup B$: We first observe that

$$\left(\frac{2}{3}\right)^6 = \frac{64}{729} \leq 10^{-1}, \quad \text{i.e.} \quad \left(\frac{2}{3}\right)^{6n} \leq 10^{-n}.$$

It is therefore possible to construct, for an arbitrary real number $\varepsilon > 0$, a positive integer j such that

$$\left(\frac{2}{3}\right)^j (\beta - \alpha) < \varepsilon.$$

As for all positive integers $n \geq j, m \geq j$ the real numbers $\alpha_n, \alpha_m, \beta_n, \beta_m$ belong to the interval $[\alpha_j; \beta_j]$, we have

$$\max(|\alpha_n - \alpha_m|, |\beta_n - \beta_m|, |\alpha_n - \beta_m|) \leq \beta_j - \alpha_j$$

and

$$\beta_j - \alpha_j = \frac{2}{3}(\beta_{j-1} - \alpha_{j-1}) = \left(\frac{2}{3}\right)^2 (\beta_{j-2} - \alpha_{j-2})$$
$$= \ldots = \left(\frac{2}{3}\right)^j (\beta - \alpha) < \varepsilon.$$

Thus the real number

$$\xi = \lim A = \lim B = \lim A \sqcup B$$

exists.
The function f is continuous on $[\alpha; \beta]$. As we have, for all positive integers n, the inequality $f(\alpha_n) \leq \eta$, we also have $f(\xi) \leq \eta$. As we have, for all positive integers n, the inequality $f(\beta_n) \geq \eta$, we also have $f(\xi) \geq \eta$. This proves $f(\xi) = \eta$.
The real number ξ is uniquely determined for the following reason: For any ξ' that is apart form ξ, we either have $\xi' < \xi$ with the consequence $f(\xi') < \eta$, or we have $\xi' > \xi$ with the consequence $f(\xi') > \eta$. ∎

Let α, β denote two real numbers such that $\alpha < \beta$ and assume that the real function f is defined on the compact interval $[\alpha; \beta]$.

If f is strictly monotone increasing, it is possible to construct a uniquely determined function g defined on $[f(\alpha); f(\beta)]$ such that the concatenations $f \circ g$ and $g \circ f$ are identical functions defined on $[f(\alpha); f(\beta)]$ and on $[\alpha; \beta]$. This inverse function g also is strictly monotone increasing.

If f is strictly monotone decreasing, it is possible to construct a uniquely determined function g defined on $[f(\beta); f(\alpha)]$ such that the concatenations $f \circ g$ and $g \circ f$ are identical functions defined on $[f(\beta); f(\alpha)]$ and on $[\alpha; \beta]$. This inverse function g also is strictly monotone decreasing.

Proof. We again only formulate the proof for strictly monotone increasing functions: The intermediate value theorem for strictly monotone functions shows how to pick up, for any $y \in [f(\alpha); f(\beta)]$, the uniquely determined $x \in [\alpha; \beta]$ such that $f(x) = y$, and we call this assignment that appoints, given the argument y, the value x of the inverse function g. The relations

$$f \circ g(y) = f(g(y)) = f(x) = y$$

for all real numbers $y \in [f(\alpha); f(\beta)]$, and

$$g \circ f(x) = g(f(x)) = g(y) = x$$

for all real numbers $x \in [\alpha; \beta]$ are consequences of this definition. The intermediate value theorem for strictly monotone functions further implies that g is uniquely determined. Suppose that $\eta' < \eta''$, and that $f(\xi') = \eta'$, $f(\xi'') = \eta''$: The arguments ξ' and ξ'' must be apart, and the possibility $\xi' > \xi''$ is excluded. Therefore we have $\xi' < \xi''$, i.e. $g(\eta') < g(\eta'')$. ∎

Let n denote a positive integer. There exists a uniquely determined strictly monotone increasing function $\sqrt[n]{\cdot}$, the *n-th root*, defined on $]0; \infty[$, such that, for positive real numbers x, y, the equation $x^n = y$ is solved by $x = \sqrt[n]{y}$.

4.4 Sequences and sets of continuous functions

4.4.1 Pointwise and uniform convergence

Let S and T denote two metric spaces, and let

$$F = (f_1, f_2, \ldots, f_n, \ldots)$$

denote a sequence consisting of functions $f_n : S \to T$ for each positive integer n. The sequence F is called *pointwise convergent* if and only if there exists a function $f : S \to T$ with the following property: for any real $\varepsilon > 0$ and any point $x \in S$ it is possible to construct a positive integer j such that, for any integer $n \geq j$, the inequality $\|f_n(x) - f(x)\| < \varepsilon$ holds.
The sequence F is called *uniformly convergent* if and only if there exists a function $f : S \to T$ with the following property: for any real $\varepsilon > 0$ it is possible to construct a positive integer k such that, for any integer $n \geq k$ and any point $x \in S$, the inequality $\|f_n(x) - f(x)\| < \varepsilon$ holds.

Using the notation above, suppose that F is pointwise convergent. Then the function f is uniquely determined, and it is called the (pointwise) limit of F with the abbreviation $f = \lim F$.

Proof. For any point $x \in S$ the sequence

$$F(x) = (f_1(x), f_2(x), \ldots, f_n(x), \ldots)$$

is convergent in T and its limit $f(x) = \lim F(x)$ is uniquely determined. ∎

Uniform convergence implies pointwise convergence.

Proof. Using the notation of the definition above, it suffices to choose the integer $j = k$. ∎

The CAUCHY-criterion of pointwise resp. uniform convergence. *Let S denote a metric space, let T denote a complete metric space, and let*

$$F = (f_1, f_2, \ldots, f_n, \ldots)$$

denote a sequence consisting of functions $f_n : S \to T$ for each positive integer n.
1. The sequence F is pointwise convergent if and only if, for any real $\varepsilon > 0$ and any point $x \in S$, it is possible to construct a positive integer l such that, for any integer $n \geq l$ and any integer $m \geq l$, the inequality $\|f_n(x) - f_m(x)\| < \varepsilon$ holds.
2. The sequence F is uniformly convergent if and only if, for any real $\varepsilon > 0$, it

is possible to construct a positive integer l such that, for any integer $n \geq l$, any integer $m \geq l$, and any point $x \in S$, the inequality $\|f_n(x) - f_m(x)\| < \varepsilon$ holds.

Proof. 1. Suppose that, for any real $\varepsilon > 0$ and any point $x \in S$, it is possible to construct a positive integer l such that, for any integer $n \geq l$ and any integer $m \geq l$, the inequality $\|f_n(x) - f_m(x)\| < \varepsilon$ holds. Then, for any point $x \in S$, the sequence

$$F(x) = (f_1(x), f_2(x), \ldots, f_n(x), \ldots)$$

fulfills the condition of the CAUCHY-criterion within the complete metric space T and therefore converges in T with $f(x) = \lim F(x)$ as its limit. This defines a function $f : S \to T$. Let $\varepsilon > 0$ denote an arbitrary real number, let $x \in S$ denote an arbitrary point, and construct a positive decimal number $e < \varepsilon$. Then it is possible to construct a positive integer j such that, for any integer $n \geq j$ and any integer $m \geq j$, the inequality

$$\|f_n(x) - f_m(x)\| < e$$

holds. The estimation of the limit implies that

$$\|f_n(x) - f(x)\| \leq e < \varepsilon$$

for any integer $n \geq j$, and this proves $f = \lim F$.

Suppose on the other hand that F is pointwise convergent with $f = \lim F$, and let $\varepsilon > 0$ denote an arbitrary real number and let $x \in S$ denote an arbitrary point. Then it is possible to construct a positive integer l such that, for any integer $n \geq l$, the inequality

$$\|f_n(x) - f(x)\| < \frac{\varepsilon}{2}$$

holds. As we also have

$$\|f_m(x) - f(x)\| < \frac{\varepsilon}{2}$$

for any integer $m \geq l$, we conclude $\|f_n(x) - f_m(x)\| < \varepsilon$ by the triangle inequality.

2. Suppose that, for any real $\varepsilon > 0$, it is possible to construct a positive integer l such that, for any integer $n \geq l$, any integer $m \geq l$, and any point $x \in S$, the inequality $\|f_n(x) - f_m(x)\| < \varepsilon$ holds. Then, as before, for any point $x \in S$ the sequence

$$F(x) = (f_1(x), f_2(x), \ldots, f_n(x), \ldots)$$

fulfills the condition of the CAUCHY-criterion within the complete metric space T and therefore converges in T with $f(x) = \lim F(x)$ as its limit. This defines a function $f : S \to T$ which, as before, proves to be the pointwise limit of F. It is also the uniform limit of F: Let $\varepsilon > 0$ denote an arbitrary real number, and construct a positive decimal number $e < \varepsilon$. Then it is possible to construct a positive integer k such that, for any integer $n \geq k$, any integer $m \geq k$, and any

point $x \in S$, the inequality $\|f_n(x) - f_m(x)\| < e$ holds. The estimation of the limit implies that
$$\|f_n(x) - f(x)\| \le e < \varepsilon$$
for any integer $n \ge j$ and any point $x \in S$.

Suppose on the other hand that F is uniformly convergent with $f = \lim F$, and let $\varepsilon > 0$ denote an arbitrary real number. Then it is possible to construct a positive integer l such that, for any integer $n \ge l$ and any point $x \in S$, the inequality
$$\|f_n(x) - f(x)\| < \frac{\varepsilon}{2}$$
holds. As we also have
$$\|f_m(x) - f(x)\| < \frac{\varepsilon}{2}$$
for any integer $m \ge l$ and any point $x \in S$, we conclude $\|f_n(x) - f_m(x)\| < \varepsilon$ by the triangle inequality. ∎

Let S denote a metric space, let T denote a complete metric space, and let $F = (f_1, f_2, \ldots, f_n, \ldots)$ denote a sequence consisting of functions $f_n : S \to T$ for each positive integer n. The sequence F is uniformly convergent if there exists a sequence $A = (\alpha_1, \alpha_2, \ldots, \alpha_n, \ldots)$ of real numbers $\alpha_1, \alpha_2, \ldots, \alpha_n, \ldots$ such that
1. A is convergent with $\lim A = 0$, and
2. for any positive integer n, any positive integer $m \ge n$, and any point $x \in S$, we have $\|f_n(x) - f_m(x)\| \le \alpha_n$.

Proof. Let $\varepsilon > 0$ denote an arbitrary real number. $\lim A = 0$ implies the existence of a positive integer k such that $\alpha_n < \varepsilon$ holds for all $n \ge k$. Then, for any positive integer $n \ge k$, any positive integer $m \ge k$, and any point $x \in S$, we have
$$\|f_n(x) - f_m(x)\| \le \max(\alpha_n, \alpha_m) < \varepsilon.$$
∎

Let S and T denote two complete metric spaces, and let U denote a subset of S. Let further ξ denote a point that belongs to the cover of U, and let $F = (f_1, f_2, \ldots, f_n, \ldots)$ denote a uniformly convergent sequence consisting of functions $f_n : U \to T$ for each positive integer n. Suppose that for each positive integer n the function f_n is continuous at ξ. Then $f = \lim F$ is continuous at ξ.

Proof. Let $\varepsilon > 0$ denote an arbitrary real number and construct a positive decimal number $e < \varepsilon$. There exists a positive integer k such that, for all integers $n \ge k$ and $m \ge k$, we have $\|f_n(x) - f_m(x)\| < e$. The continuity of the functions f_n and f_m at ξ implies that
$$\|f_n(\xi) - f_m(\xi)\| \le e < \varepsilon,$$
and this proves the convergence of the sequence
$$F(\xi) = (f_1(\xi), f_2(\xi), \ldots, f_n(\xi), \ldots).$$

We now use the inequality

$$\|f(x) - \lim F(\xi)\| \leq \|f(x) - f_n(x)\| \\ + \|f_n(x) - f_n(\xi)\| + \|f_n(\xi) - \lim F(\xi)\|$$

which holds for any positive integer n and any point $x \in U$. It is possible to construct a positive integer j_1 such that

$$\|f_n(\xi) - \lim F(\xi)\| < \frac{\varepsilon}{3}$$

for all integers $n \geq j_1$. It is also possible to construct a positive integer j_2 such that

$$\|f(x) - f_n(x)\| < \frac{\varepsilon}{3}$$

for all integers $n \geq j_2$ and all points $x \in U$. We now set $n = \max(j_1, j_2)$. The continuity of f_n at ξ allows us to construct a real number $\delta > 0$ such that, for all $x \in U$, the inequality $\|x - \xi\| < \delta$ implies that

$$\|f_n(x) - f_n(\xi)\| < \frac{\varepsilon}{3}.$$

Thus we have

$$\|f(x) - \lim F(\xi)\| < \varepsilon$$

for all $x \in U$ fulfilling $\|x - \xi\| < \delta$. ∎

4.4.2 Sequences of functions defined on compact spaces

Consider the following example: For any positive integer n let $f_n : [0; 1] \to \mathbb{R}$ denote the function defined by $f_n(x) = x^n$.
1. Suppose we have $0 \leq x < 1$. In this case the sequence

$$X = \left(x, x^2, \ldots, x^n, \ldots\right)$$

converges to zero. A possible proof of this fact relies on the *formula of the geometric sum*

$$a^n + a^{n-1}b + a^{n-2}b^2 \ldots + a^2 b^{n-2} + ab^{n-1} + b^n = \frac{a^{n+1} - b^{n+1}}{a - b}$$

for any real numbers a, b with $a \neq b$. This formula immediately follows by induction or by multiplication of both sides with $a - b$. In the special case $a = 1$ and $b = x$, BERNOULLI concluded the inequality

$$(n+1)x^n \leq 1 + x + x^2 + \ldots + x^{n-2} + x^{n-1} + x^n = \frac{1 - x^{n+1}}{1 - x} \leq \frac{1}{1 - x}$$

with the consequence

$$x^n \leq \frac{1}{n+1} \frac{1}{1-x}.$$

Let $\varepsilon > 0$ denote an arbitrary real number. Construct the positive integer j so big that $j \geq 1/(1-x)\varepsilon$ holds. Then we have for all integers $n \geq j$

$$|x_n - 0| = x_n \leq \frac{1}{n+1}\frac{1}{1-x} < (1-x)\varepsilon\frac{1}{1-x} = \varepsilon,$$

i.e. $\lim X = 0$.

2. Suppose we have $x = 1$. In this case the sequence $X = (x, x^2, \ldots, x^n, \ldots)$ obviously converges to one.

Nevertheless, it is forbidden to assert that $F = (f_1, f_2, \ldots, f_n, \ldots)$ converged pointwise to a function $f : [0; 1] \to \mathbb{R}$ fulfilling $f(x) = 0$ if $0 \leq x < 1$ and $f(1) = 1$. The reason is: it is impossible to decide for any given real number x between the alternatives $x < 1$ or $x \geq 1$. The only conclusion we can draw is the following:

Let q denote a positive real number with $q < 1$. For any positive integer n let $f_n : [0; q] \to \mathbb{R}$ denote the function defined by $f_n(x) = x^n$. Then the sequence $F = (f_1, f_2, \ldots, f_n, \ldots)$ converges pointwise to zero. It even converges uniformly because

$$\lim \left(q, q^2, \ldots, q^n, \ldots \right) = 0$$

and

$$|f_n(x) - f_m(x)| = |x^n - x^m| \leq x^n \leq q^n$$

for all integers n and $m \geq n$ and all $x \in [0; q]$.

This example is paradigmatic for the following perception:

Theorem of DINI **and** BROUWER. *Let S denote a compact metric space, let T be a complete metric space, and let $F = (f_1, f_2, \ldots, f_n, \ldots)$ denote a pointwise converging sequence consisting of functions $f_n : S \to T$ for each positive integer n. Then F is a uniformly converging sequence.*

Proof. We define $f = \lim F$. For any real $\varepsilon > 0$ and any point $x \in S$ it is possible to construct a positive integer j such that, for any integer $n \geq j$, the inequality $\|f_n(x) - f(x)\| < \varepsilon$ holds. According to the theorem of HEINE and BOREL it is possible to construct a positive integer k such that, for all points $x \in S$ and all integers $n \geq k$, the inequality $\|f_n(x) - f(x)\| < \varepsilon$ holds. This proves the asserted uniform convergence. ∎

In the original version of the theorem, DINI presupposed a monotony-condition on the sequence F and the continuity of all functions f_n plus the continuity of $f = \lim F$ to achieve uniform convergence. Within BROUWER's intuitionistic mathematics, all these additional conditions are superfluous.

4.4.3 Spaces of functions defined on compact spaces

Let S denote a compact metric space and let T denote a complete metric space. The set $C(S, T)$ of all functions $f : S \to T$ becomes a metric space by defining

$$\|f - g\|_\infty = \sup \|f - g\|(S)$$

as distance between the functions $f : S \to T$ and $g : S \to T$. The function $\|f - g\| : S \to \mathbb{R}$ thereby is defined by the formula

$$\|f - g\|(x) = \|f(x) - g(x)\|,$$

where $\|f(x) - g(x)\|$ is the distance between the points $f(x)$ and $g(x)$ in T. The metric space $\mathcal{C}(S, T)$ is complete.

Proof. The positivity of $\|f - g\|_\infty$ is obvious. Suppose $\|f - g\|_\infty = 0$, then we have $\|f(x) - g(x)\| = 0$ for all $x \in S$, therefore $f = g$. If there exists a point x_0 such that $f(x_0) \neq g(x_0)$, then the inequality $\|f - g\|_\infty > 0$ follows immediately and vice-versa. Thus the *apartness* of two functions $f, g \in \mathcal{C}(S, T)$ is well defined by the statement that there exists a point $x_0 \in S$ with the property $f(x_0) \neq g(x_0)$. Suppose that f, g, h are three functions of $\mathcal{C}(S)$. As we have, by the triangle inequality within T,

$$\|f(x) - h(x)\| - \|g(x) - h(x)\| \leq \|f(x) - g(x)\|$$

for all $x \in S$, we can derive

$$\|f(x) - h(x)\| - \|g - h\|_\infty \leq \|f(x) - g(x)\|$$

for all $x \in S$, with the conclusion

$$\|f - h\|_\infty - \|g - h\|_\infty \leq \|f - g\|_\infty.$$

Suppose that $F = (f_1, f_2, \ldots, f_n, \ldots)$ denotes a fundamental sequence in the space $\mathcal{C}(S, T)$. Then it is possible to construct, for any real $\varepsilon > 0$, a positive integer j such that, for all $m \geq j$ and all $n \geq j$, the inequality $\|f_n - f_m\|_\infty < \varepsilon$ holds. This implies $\|f_n(x) - f_m(x)\| < \varepsilon$ for all points $x \in S$ and therefore the uniform convergence of F. The CAUCHY-criterion of uniform convergence confirms the existence of a function $f \in \mathcal{C}(S, T)$ with $\lim F = f$. Let $\varepsilon > 0$ denote an arbitrary real number. There exists a positive decimal number $e < \varepsilon$ and a positive integer k such that, for all $n \geq k$, the inequality $\|f_n(x) - f(x)\| < e$ holds for all points $x \in S$. This implies $\|f_n - f\|_\infty \leq e < \varepsilon$ for all $n \geq k$. In other words: the uniform convergence is equivalent to the convergence within $\mathcal{C}(S, T)$, and the CAUCHY-criterion of uniform convergence proves that $\mathcal{C}(S, T)$ is complete. ∎

In the case $T = \mathbb{R}$, we simply write $\mathcal{C}(S)$ instead of $\mathcal{C}(S, \mathbb{R})$.

Suppose that $\mathcal{U}$ is a subset of $\mathcal{C}(S)$ and that $x_1, x_2, \ldots, x_n$ are n points in S. We define $\mathcal{U}(x_1, \ldots, x_n)$ to be the set consisting of all points $(v_1, \ldots, v_n)$ in $\mathbb{R}^n$ with the following property: there exists a function $f \in \mathcal{U}$ such that

$$f(x_1) = v_1, \quad f(x_2) = v_2, \quad \ldots, \quad f(x_n) = v_n.$$

We further define $\mathcal{U}$ to be *equicontinuous* if and only if for any real $\varepsilon > 0$ there exists a real $\delta > 0$ such that for any two points $x' \in S$, $x'' \in S$ and any function $f \in \mathcal{U}$ the inequality $\|x' - x''\| < \delta$ implies $|f(x') - f(x'')| < \varepsilon$.

Theorem of ARZELA **and** ASCOLI. *Let $\mathcal{U}$ be an equicontinuous subset of $\mathcal{C}(S)$ with the additional property that for each real $\varepsilon > 0$ there exists a finite ε-net $(x_1, x_2, \ldots, x_n)$ of S for which $\mathcal{U}(x_1, \ldots, x_n)$ is totally bounded. Then $\mathcal{U}$ itself is totally bounded.*

Proof. Let $\varepsilon > 0$ denote an arbitrary real number. By definition of equicontinuity, there exists a real $\delta > 0$ such that for any two points $x' \in S$, $x'' \in S$ and any function $f \in \mathcal{U}$ the inequality $\|x' - x''\| < \delta$ implies $|f(x') - f(x'')| < \varepsilon/3$. Further, there exists a δ-net $(x_1, x_2, \ldots, x_n)$ of S for which $\mathcal{U}(x_1, \ldots, x_n)$ is totally bounded. We therefore can exhibit functions $f_1, f_2, \ldots, f_m$ from $\mathcal{U}$ such that the points

$$(f_1(x_1), \ldots, f_1(x_n)), (f_2(x_1), \ldots, f_2(x_n)), \ldots, (f_m(x_1), \ldots, f_m(x_n))$$

form an $\varepsilon/4$-net of $\mathcal{U}(x_1, \ldots, x_n)$. This means that for an arbitrary $f \in \mathcal{U}$ there exists a positive integer $j \leq m$ with the property

$$\max\left(|f_j(x_1) - f(x_1)|, \ldots, |f_j(x_n) - f(x_n)|\right) < \frac{\varepsilon}{4}.$$

For an arbitrary $x \in S$ there exists a positive integer $k \leq n$ with the property $\|x - x_k\| < \delta$. Thus we have

$$\begin{aligned} |f_j(x) - f(x)| &\leq |f_j(x) - f_j(x_k)| + |f_j(x_k) - f(x_k)| \\ &\quad + |f(x_k) - f(x)| \\ &\leq \frac{\varepsilon}{3} + \frac{\varepsilon}{4} + \frac{\varepsilon}{3} = \frac{11\varepsilon}{12}. \end{aligned}$$

Since x is an arbitrary point of S, it follows that $\|f_j - f\|_\infty < \varepsilon$. Thus the functions $f_1, f_2, \ldots, f_m$ form a finite ε-net of $\mathcal{U}$. ∎

4.4.4 Compact spaces of functions

The theorem of ARZELA and ASCOLI allows to detect an important example of a compact subspace $\mathcal{U}$ of $\mathcal{C}(S)$: We call a strictly monotone increasing function $h : [0; \infty[\to [0; \infty[$ a HOELDER-*function* if and only if

$$h(0) = 0 \quad \text{and} \quad h(\alpha + \beta) \leq h(\alpha) + h(\beta)$$

for all real numbers $\alpha \geq 0$, $\beta \geq 0$.

Let S denote a compact metric space, let h denote a HOELDER-*function*, and let $\mathcal{U}$ be a bounded subset of $\mathcal{C}(S)$ with the property that

$$|f(x) - f(y)| \leq h(\|x - y\|)$$

for all functions $f \in \mathcal{U}$ and all points x, y in S. Then $\mathcal{U}$ is compact.

Proof. The fact that $\mathcal{U}$ is bounded is equivalent to the existence of a real number μ such that $\|f\|_\infty \leq \mu$. (In this formula $\|f\|_\infty$ abbreviates $\|f - 0\|_\infty$ and 0 stands for the constant function with value zero.)

Suppose that F is a convergent sequence of functions $f_1, f_2, \ldots, f_n, \ldots$ in $\mathcal{U}$ and $f = \lim F$. The inequalities $\|f_n\|_\infty \leq \mu$ for all positive integers n imply $\|f\|_\infty \leq \mu$. Further, the inequalities

$$|f_n(x) - f_n(y)| \leq h(\|x - y\|)$$

for all points x, y in S and all positive integers n imply

$$|f(x) - f(y)| \leq h(\|x - y\|)$$

for all points x, y in S. Therefore $\mathcal{U}$ is a closed subset of $\mathcal{C}(S)$, and it is enough to show that $\mathcal{U}$ is totally bounded.

Suppose that $\varepsilon > 0$ denotes an arbitrary real number. Since the HOELDER-function h is strictly monotone increasing, the real number $\delta > 0$ is well defined by the equation $h(\delta) = \varepsilon$. Further, for all points x, y in S the inequality $\|x - y\| < \delta$ implies

$$|f(x) - f(y)| \leq h(\|x - y\|) < h(\delta) = \varepsilon.$$

Therefore $\mathcal{U}$ is an equicontinuous subset of $\mathcal{C}(S)$.

By the theorem of ARZELA and ASCOLI it will suffice to show that for each real $\varepsilon > 0$ there exists a finite ε-net $(x_1, x_2, \ldots, x_n)$ of S for which $\mathcal{U}(x_1, \ldots, x_n)$ is totally bounded.

First we show the following: For each real $\varepsilon > 0$ there exists a finite ε-net $(x_1, x_2, \ldots, x_n)$ of S that is *entirely discrete*. This means that for all positive integers j, k the inequality $j \neq k$ implies the apartness $x_j \neq x_k$. We construct a positive decimal number $e < \varepsilon$ and a finite e-net $(y_1, y_2, \ldots, y_m)$ of S. By the dichotomy lemma, for each pair of integers j, k with $1 \leq j < k \leq m$ at least one of the two inequalities

$$\|y_j - y_k\| < \frac{\varepsilon - e}{m - 1} \quad \text{or} \quad \|y_j - y_k\| > \frac{\varepsilon - e}{m}$$

must hold. Given the pair (j, k) of integers j, k with $1 \leq j < k \leq m$, we eliminate y_k from the e-net $(y_1, y_2, \ldots, y_m)$ if the equality

$$\|y_j - y_k\| < \frac{\varepsilon - e}{m - 1}$$

holds. This systematic elimination produces a sequence $(x_1, x_2, \ldots, x_n)$ which is a subsequence of $(y_1, y_2, \ldots, y_m)$ and which, by construction, is entirely discrete. Let x denote an arbitrary point of S. There exists a point y_k such that the inequality $\|x - y_k\| < e$ holds. If y_k is not eliminated, i.e. if $y_k = x_l$, then we have

$$\|x - x_l\| < \varepsilon.$$

If y_k is eliminated, there exists a point y_j such that $j < k$ and
$$\|y_j - y_k\| < \frac{\varepsilon - e}{m-1}.$$
If y_j is not eliminated, i.e. if $y_j = x_l$, then we have
$$\begin{aligned}\|x - x_l\| &\leq \|x - y_k\| + \|y_j - y_k\| \\ &< e + \frac{\varepsilon - e}{m-1} \leq \varepsilon.\end{aligned}$$
If y_j is eliminated, the argument above can be repeated – but this at most $m-1$ times. Finally we gain a point $y_r = x_l$ for which
$$\begin{aligned}\|x - x_l\| &\leq \|x - y_k\| + \ldots + \|y_r - y_s\| \\ &< e + (m-1)\frac{\varepsilon - e}{m-1} \leq \varepsilon.\end{aligned}$$

Next we define for integers j, k with $1 \leq j \leq n$ and $1 \leq k \leq n$
$$\rho_{jk} = \begin{cases} h(\|x_j - x_k\|) & \text{if } j \neq k, \\ 1 & \text{if } j = k. \end{cases}$$
and we prove that $\mathcal{U}(x_1, \ldots, x_n)$ coincides with the set $V \subseteq \mathbb{R}^n$ consisting of all points $v = (v_1, \ldots, v_n)$ fulfilling
$$|v_j| \leq \mu \quad \text{and} \quad |v_j - v_k| \leq \rho_{jk}$$
for all integers j, k with $1 \leq j < k \leq n$. On the one hand we obviously have $\mathcal{U}(x_1, \ldots, x_n) \subseteq V$. Therefore it is enough to show $V \subseteq \mathcal{U}(x_1, \ldots, x_n)$. Since S is totally bounded, $(x_1, \ldots, x_n)$ can be extended to an infinite sequence $(x_1, x_2, \ldots, x_m, \ldots)$ of points of S which is dense in S. Let us consider a point $v = (v_1, \ldots, v_n)$ in V. We continue the finite sequence $(v_1, \ldots, v_n)$ to an infinite sequence $(v_1, v_2, \ldots, v_m, \ldots)$ such that, for all positive integers j, k,
$$|v_j| \leq \mu \quad \text{and} \quad |v_j - v_k| \leq h(\|x_j - x_k\|).$$
This is done inductively. Certainly, the condition above holds for all positive integers j, k with $j \leq n$ and $k \leq n$. Assume that $v_1, v_2, \ldots, v_m$ have already been constructed to satisfy the condition above. The hypotheses of the HOELDER-function h ensures that
$$\begin{aligned}|v_j - v_k| &\leq h(\|x_j - x_k\|) \leq h(\|x_j - x_{m+1}\| + \|x_{m+1} - x_k\|) \\ &\leq h(\|x_j - x_{m+1}\|) + h(\|x_{m+1} - x_k\|),\end{aligned}$$
and a fortiori
$$v_j - h(\|x_j - x_{m+1}\|) \leq v_k + h(\|x_{m+1} - x_k\|)$$

for all positive integers j, k with $j \leq m$ and $k \leq m$. This implies for the numbers

$$\phi = \max\left(v_1 - h\left(\|x_1 - x_{m+1}\|\right), \ldots, v_m - h\left(\|x_m - x_{m+1}\|\right)\right)$$

and

$$\psi = \min\left(v_1 + h\left(\|x_1 - x_{m+1}\|\right), \ldots, v_m + h\left(\|x_m - x_{m+1}\|\right)\right)$$

the inequality $\phi \leq \psi$. If we define

$$v_{m+1} = \frac{\phi + \psi}{2},$$

then we have $\phi \leq v_{m+1} \leq \psi$ and therefore

$$|v_j - v_{m+1}| \leq h\left(\|x_j - x_{m+1}\|\right)$$

for all integers $j \leq m + 1$. Thus the sequence $(v_1, v_2, \ldots, v_n, \ldots)$ is constructed inductively. The assignment f is defined by the formula $f(x_j) = v_j$ for all positive integers j. By construction, f is continuous at each point ξ of S: Suppose $\varepsilon > 0$ denotes an arbitrary real number. Then there exists a real number $\delta > 0$ such that $h(2\delta) \leq \varepsilon$, and the inequalities

$$\|x_j - \xi\| < \delta \quad \text{and} \quad \|x_k - \xi\| < \delta$$

imply

$$\begin{aligned} |f(x_j) - f(x_k)| &= |v_j - v_k| \leq h\left(\|x_j - x_k\|\right) \\ &\leq h\left(\|x_j - \xi\| + \|x_k - \xi\|\right) < h(2\delta) \leq \varepsilon. \end{aligned}$$

We therefore have $f \in \mathcal{U}$. Since $v = (f(x_1), \ldots, f(x_n))$, the point v belongs to $\mathcal{U}(x_1, \ldots, x_n)$.

Finally, we look at this set $V = \mathcal{U}(x_1, \ldots, x_n)$ of points $v = (v_1, \ldots, v_n)$ in $\mathbb{R}^n$ fulfilling

$$|v_j| \leq \mu \quad \text{and} \quad |v_j - v_k| \leq \rho_{jk}$$

for all integers j, k with $1 \leq j < k \leq n$. We define $[\alpha; \beta] \subseteq \mathbb{R}^n$ with

$$\alpha = (-\mu, \ldots, -\mu) \quad \text{and} \quad \beta = (\mu, \ldots, \mu),$$

and we define the function

$$g: [\alpha; \beta] \to [\alpha; \beta]$$

by $g(u_1, \ldots, u_n) = (v_1, \ldots, v_n)$ with

$$v_l = \frac{u_l}{\max\left(1, \dfrac{|u_1 - u_2|}{\rho_{12}}, \dfrac{|u_1 - u_3|}{\rho_{13}}, \ldots, \dfrac{|u_j - u_k|}{\rho_{jk}}, \ldots, \dfrac{|u_{n-1} - u_n|}{\rho_{n-1,n}}\right)}$$

for all positive integers $l \leq n$. We even have

$$g : [\alpha; \beta] \to V,$$

and, since $g(v_1, \ldots, v_n) = (v_1, \ldots, v_n)$ for all $(v_1, \ldots, v_n) \in V$, the set

$$\mathcal{U}(x_1, \ldots, x_n) = V = g([\alpha; \beta])$$

is totally bounded. ■

Let λ be a positive real number. The function h defined by $h(x) = \lambda x$ obviously is a HOELDER-function. Thus we can formulate the following corollary:

Let S denote a compact metric space, and let λ, μ be positive real numbers. Let $\mathcal{U}$ be the set of all f in $C(S)$ with $\|f\|_\infty \leq \mu$ which satisfy the so-called LIPSCHITZ-*condition*

$$|f(x) - f(y)| \leq \lambda \|x - y\|$$

for all points x, y of S. Then $\mathcal{U}$ is compact.

5
Literature

M. van Atten, D. van Dalen, R. Tieszen: Brouwer and Weyl: The Phenomenology and Mathematics of the Intuitive Continuum, *Philosophia Mathematica* 10, 203-226 (2002)

P. Benacerraf, H. Putnam (editors): *Philosophy of Mathematics. Selected Readings.* Cambridge University Press, Cambridge, 1983

O. Becker: *Größe und Grenze der mathematischen Denkweise.* K. Alber, Freiburg, 1959

M. Beeson: *Foundations of Constructive Mathematics.* Springer, Berlin, Heidelberg, New York, 1985

J. E. Bell: *A Primer of Infinitesimal Analysis.* Cambridge University Press, Cambridge, 1998

J. E. Bell: Hermann Weyl on Intuition and the Continuum. *Philosophia Mathematica* (3), 8, 2000

J. E. Bell: The Continuum in Smooth Infinitesimal Analysis. In: U. Berger, H. Osswald, P. Schuster (editors): *Reuniting the Antipodes. Constructive and Nonstandard Views of the Continuum.* Kluwer, Dordrecht, 2001

E. W. Beth: *Mathematical Thought. An Introduction to the Philosophy of Mathematics.* D. Reidel, Dordrecht, 1965

E. Bishop: *Foundations of Constructive Analysis.* McGraw-Hill, New York, 1967

E. Bishop: *Aspects of Constructivism.* Las Cruces. New Mexico State University, 1972

E. Bishop: The Crisis in Contemporary Mathematics. *Historia Mathematica* 2, 507-517 (1975)

E. Bishop, D. S. Bridges: *Constructive Analysis.* Springer, Berlin, Heidelberg, New York, 1985

B. Bolzano: *Paradoxien des Unendlichen.* Reclam, Leipzig, 1851

É. Borel: *L'imaginaire et le réel en mathématiques et en physique.* Alvin Michel, Paris, 1952

D. S. Bridges: Some Notes on Continuity in Constructive Analysis. *Bulletin of the London Mathematical Society* 8, 179-182 (1976)

D. S. Bridges: A Criterion for Compactness in Metric Spaces. *Zeitschrift für mathematische Logik und Grundlagen der Mathematik* 25, 97-98 (1979)

D. S. Bridges, R. Mines: What is Constructive Mathematics? *The Mathematical Intelligencer* 6(4), 32-38 (1985)

D. S. Bridges, R. Richman: *Varieties of Constructive Mathematics.* Cambridge University Press, Cambridge, 1987

D. S. Bridges: A General Constructive Intermediate Value Theorem. *Zeitschrift für mathematische Logik und Grundlagen der Mathematik* 35, 433-435 (1989)

D. S. Bridges: *Foundations of Real and Abstract Analysis.* Springer, Berlin, Heidelberg, New York, 1998

D. S. Bridges: Sequential, Pointwise, and Uniform Continuity: A Constructive Note. *Mathematical Logic Quarterly* 39, 55-61 (1993)

D. S. Bridges: A Constructive Look at the Real Number Line. In: P. Ehrlich (editor): *Synthèse: Real Numbers, Generalizations of the Reals and Theories of Continua.* Kluwer Academic Publishers, Amsterdam, 1994

D. S. Bridges, H. Ishihara, P. Schuster: Sequential Compactness in Constructive Analysis. *Sitzungsberichte der Österreichischen Akademie der Wissenschaften* II 208, 159-163 (1999)

D. S. Bridges, P. Schuster, Luminiţa Vîţă: Strong versus Uniform Continuity, a Constructive Round. *Quaestiones Mathematicae* 26, 171.190 (2003)

J. R. Brown: *Philosophy of Mathematics. An Introduction to the World of Proofs and Pictures.* Routledge, London, 1999

L. E. J. Brouwer: Intuitionism and Formalism. *Bulletin of the American Mathematical Society* 20, 81-96 (1912)

L. E. J. Brouwer: Besitzt jede reelle Zahl eine Dezimalbruchentwicklung? *Mathematische Annalen* 83, 201-210 (1922)

L. E. J. Brouwer: Beweis, dass jede volle Funktion gleichmäßig stetig ist. *Proceedings of the Koninklijke Akademie van Wetenschappen* 27, 189-193 (1924)

L. E. J. Brouwer: Zur Begründung der intuitionistischen Mathematik I. *Mathematische Annalen* 93, 244-257 (1925)

L. E. J. Brouwer: Zur Begründung der intuitionistischen Mathematik II. *Mathematische Annalen* 95, 453-472 (1926)

L. E. J. Brouwer: Zur Begründung der intuitionistischen Mathematik III. *Mathematische Annalen* 96, 451-488 (1927)

L. E. J. Brouwer: Über Definitionsbereiche von Funktionen. *Mathematische Annalen* 97, 60-75 (1927)

L. E. J. Brouwer: Intuitionistische Betrachtungen über den Formalismus, *Sitzungsberichte der Preußischen Akademie der Wissenschaften* 48-52 (1928)

L. E. J. Brouwer: Mathematik, Wissenschaft und Sprache. *Monatshefte für Mathematik und Physik* 36, 153-164 (1929)

L. E. J. Brouwer: *The Cambridge Lectures;* edited by D. van Dalen. Cambridge University Press, Cambridge, 1981

L. E. J. Brouwer: *Intuitionismus;* herausgegeben von D. van Dalen. B.I. Wissenschaftsverlag, Mannheim, 1992

R. Courant, H. Robbins: *What Is Mathematics?* Oxford University Press, Oxford, 1980

D. van Dalen, A. S. Troelstra: *Constructivism in Mathematics I, II.* North Holland Publishing Company, Amsterdam, 1988

D. van Dalen: Brouwer: The Genesis of His Intuitionism. *Dialectica* 32, 291-303 (1978)

D. van Dalen: Braucht die konstruktive Mathematik Grundlagen? *Jahresberichte der Deutschen Mathematikervereinigung* 84, 57-78 (1982)

D. van Dalen: Infinitesimals and the Continuity of All Functions. *Nieuw Archif for Wiskunde* 6, 191-202 (1987)

D. van Dalen: The War of the Frogs and the Mice, or the Crisis of the Mathematische Annalen. *The Mathematical Intelligencer* 12, 17-31 (1990)

D. van Dalen: Hermann Weyl's Intuitionistic Mathematics. *Bulletin of Symbolic Logic* 1, 145-169 (1995)

D. van Dalen: Why Constructive Mathematics? In: F. Stadler, W. DePauli-Schimanovich, E. Köhler (editors): *The Foundational Debate. Complexity and Constructivity in Mathematics and Physics,* Kluwer, Dordrecht, 1995, pages 141-158

D. van Dalen: How Connected is the Intuitionistic Continuum? *Journal of Symbolic Logic* 62, 1174-1150 (1997)

D. van Dalen: *Mystic, Geometer, and Intuitionist: The Life of L.E.J. Brouwer. Volume I. The Dawning Revolution.* Oxford University Press, Oxford, 1999

P. Davis, R. Hersh: *The Mathematical Experience.* Birkhäuser, Boston, 1981

R, Dedekind: *Was sind und was sollen die Zahlen?* Vieweg, Braunschweig, 1888

M. Dummet: *Elements of Intuitionism.* Clarendon Press, Oxford, 1977

M. Dummett: The Philosophy of Mathematics. In: A.C. Grayling (editor): *Philosophy 2. Further Through the Subject.* Oxford University Press, Oxford, 1995, pages 122-196

G. H. Hardy, E. M. Wright: *An Introduction to the Theory of Numbers.* Oxford University Press, Oxford, 1938

A. Heyting: *Intuitionism. An Introduction.* North Holland Publishing Company, Amsterdam, 1987

D. Hilbert: Über das Unendliche. *Mathematische Annalen* 95, 161–190 (1926)

D. Klaua: *Konstruktive Analysis.* Deutscher Verlag der Wissenschaften, Berlin, 1961

S. C. Kleene, R. E. Vesley: *The Foundations of Intuitionistic Mathematics.* North Holland Publishing Company, Amsterdam 1965

M. Kline: *Mathematics. The Loss of Certainty.* Oxford University Press, Oxford, 1980

G. Kreisel: Lawless Sequences of Natural Numbers. *Compositio Mathematica* 20, 222-248 (1968)

G. Kreisel, A. S. Troelstra: Formal Systems for Some Branches of Intuitionistic Analysis. *Annals of Mathematical Logic* 1, 229-387 (1970)

I. Lakatos: *Proofs and Refutations. The Logic of Mathematical Discovery.* Cambridge University Press, Cambridge, 1976

P. Martin-Löf: *Notes on Constructive Mathematics.* Almqvist & Wiksell, Stockholm, 1970

R. Mines, F. Richman, W. Ruitenberg: *A Course in Constructive Algebra.* Springer, Berlin, Heidelberg, New York, 1988

P. Mancoso: *From Brouwer to Hilbert: The Debate on the Foundations of Mathematics in the 1920s.* Clarendon Press, Oxford, 1998

I. Niven, H. S. Zuckerman: *An Introduction to the Theory of Numbers.* John Wiley, New York, 1960

P. Lorenzen: Das Aktual-Unendliche in der Mathematik. *Philosophia naturalis* 4: 3-11 (1957)

P. Lorenzen: *Differential und Integral.* Akademische Verlagsgesellschaft, Frankfurt am Main, 1965

P. Lorenzen: *Metamathematik.* B.I. Wissenschaftsverlag, Mannheim, 1962

P. Maddy: *Realism in Mathematics,* Clarendon Press, Oxford, 1990

J. von Neumann: The Mathematician. In: R. B. Heywood (editor): *The Works of the Mind.* Chicago, 1947, pages 180-196

A. Rényi: *Dialoge über Mathematik.* Birkhäuser, Basel, 1967

F. Richman: Meaning and Information in Constructive Mathematics. *The American Mathematical Monthly* 89, 385-388 (1982)

F. Richman (editor): *Constructive Mathematics. Lecture Notes in Mathematics 873.* Springer, Berlin, Heidelberg, New York, 1981

R. Rucker: *Infinity and the Mind. The Science and Philosophy of the Infinite.* Birkhäuser, Boston, 1982

P. Schuster: A Constructive Look at Generalized Cauchy Reals. *Mathematical Logic Quarterly* 46, 125-134 (200)

P. Schuster: Real Numbers as Black Boxes. *New Zealand Journal of Mathematics* 31, 189-202 (2002)

S. Shapiro: *Thinking About Mathematics. The Philosophy of Mathematics.* Oxford University Press, Oxford, 2000

G. Stolzenberg: Review of "Foundations of Constructive Analysis". *Bulletin of the American Mathematical Society* 76, 301-323 (1970)

G. Stolzenberg: Kann die Untersuchung der Grundlagen der Mathematik uns etwas über das Denken verraten? In: P. Watzlawick (Herausgeber): *Die erfundene Wirklichkeit.* Piper, München, 1984

R. Taschner: Constructive Mathematics for Beginners. In: *Proceedings of the International Conference on the Teaching of Mathematics.* Wiley, New York, 1998

R. Taschner: Mathematik, Logik, Wirklichkeit. Mit Kritik von: G. Asser, J. Cigler, D. van Dalen, H.-D. Ebbinghaus, U. Felgner, C. Fermüller, L. E. Fleischhacker, Y. Gauthier, D. Gernert, K. Gloede, M. Goldstern, B. J. Gut, H. Hrachovec, M. Junker, W. Kolaczia, P. H. Krauss, D. Laugwitz, A. Locker, B. Löwe, J. Maaß, H. Mehrtens, G. H. Moore, T. Mormann, F. Mühlhölzer, W. Pohlers, K. Radbruch, S. Rahman, H. Rückert, D. D. Spalt, C. Thiel, R. A. Treumann, G. Vollmer, P. Zahn. In: *EuS* 9, 425-499 (1998)

R. Taschner: Real Numbers and Functions Exhibited in Dialogues. In: U. Berger, H. Osswald, P. Schuster (editors): *Reuniting the Antipodes. Constructive and Nonstandard Views of the Continuum.* Kluwer, Dordrecht, 2001

R. Taschner: Hermeneutik der Mathematik. Über das Verstehen von Zahlen und Funktionen. *Facta Philosophica* 3, 31-57 (2001)

C. Thiel (Herausgeber): *Erkenntnistheoretische Grundlagen der Mathematik.* Gerstenberg, Hildesheim, 1982

A. S. Troelstra: *Principles of Intuitionism. Lecture Notes in Mathematics 95.* Springer, Berlin, Heidelberg, New York, 1969

5. Literature

H. Wang: *From Mathematics to Philosophy.* Routledge and Kegan, London, 1974

H. Weyl: *Das Kontinuum.* Veit & Co., Leipzig, 1918

H. Weyl: Der circulus vitiosus in der heutigen Begründung der Analysis. *Jahresbericht der Deutschen Mathematikervereinigung* 28, 85-92 (1919)

H. Weyl: Über die neue Grundlagenkrise der Mathematik. *Mathematische Zeitschrift* 10, 39-79 (1921)

H. Weyl: Die heutige Erkentnislage in der Mathematik. *Symposion* 1, 1-32 (1925)

H. Weyl: *Philosophie der Mathematik und Naturwissenschaft.* R. Oldenbourg, München, 1926

H. Weyl: Diskussionsbemerkungen zu dem zweiten Hilbertschen Vortrag über die Grundlagen der Mathematik. *Abhandlungen aus dem mathematischen Seminar der Hamburgischen Universität* 6, 86-88 (1928)

H. Weyl: Consistency in Mathematics. *The Rice Institute Pamphlet* 16, 245-265 (1929)

H. Weyl: *Die Stufen des Unendlichen.* G. Fischer, Jena, 1931

H. Weyl: *The Open World.* H. Milford, London, 1932

H. Weyl: *Mind and Nature.* University of Pennsylvania Press, Philadelphia; Oxford University Press, London, 1934

H. Weyl: *Philosophy of Mathematics and Natural Science.* Princeton University Press, 1949

L. Wittgenstein: *Bemerkungen über die Grundlagen der Mathematik – Remarks on The Foundations of Mathematics.* Basil Blackwell, Oxford, 1956

Index

absolute difference, 26
absolute value, 111
addition, 22, 111
adequate pair, 54
Anaxagoras, 14
apartness, 38, 49, 57, 123
approximation lemma, 35, 53, 55
argument, 95
Aristotle, 105

bar, 74
bar-theorem, 76
Bernoulli, Jakob, 121
Bishop, Errett, 15, 70
black-box-number, 25
Borwein, Jonathan, 19
bounded interval, 89
bounded sequence, 64
bounded set, 71
Bridges, Douglas S., 15
Brouwer, Luitzen Egbertus Jan, 13–15, 17–19, 71, 78, 122

Cantor, Georg, 1, 10, 12, 17–19, 70, 105

Cauchy criterion, 46
    criterion of Cauchy, 118
Cauchy sequence, 19
cell, 89
closed interval, 89
closed set, 78
compact interval, 92
compact metric space, 72
complete metric space, 60
concatenation, 96
connection, 84
constant function, 96
constant sequence, 52
continued fraction, 6
continuity, 97
convergence, 41, 58
coordinate, 85
cover, 78
criterion of apartness, 38
criterion of equality, 39
criterion of strict order, 27, 39
criterion of weak order, 28, 39
cut, 10
cycle, 84

decimal number, 22
Dedekind cut, 10
Dedekind, Richard, 1, 9, 10, 18, 19, 70, 105
Denjoy integral, 17
dense sequence, 82
dense set, 82
dichotomy lemma, 36
difference, 26
dimension, 89
Dini, Ulisse, 122
Dirichlet function, 16
Dirichlet's principle, 13
discrete points, 89
distance, 49
division, 22, 114

embedded function, 96
enlarged function, 96
entirely discrete sequence, 96
enumerable set, 96
equality, 22, 38, 49, 57, 96
equicontinuity, 123
equivalent metric, 86
estimation of the limit, 42
extensionality, 95
exterior, 81

Farey fraction, 1
Farey table, 2
Farey, John, 1, 2
function, 95
function defined as a pair of sequences, 96
fundamental sequence, 51

Gödel, Kurt, 18
geometric sum, 121

Hölder-function, 124
half-open interval, 89
Henstock integral, 16
Henstock, Ralph, 16
Hilbert, David, 13
Hippasos of Metapont, 3

identical function, 96
indirect proof, 27, 38
infimum, 67, 71, 91
infinite decimal number, 24
infinite sequence, 41
inner point, 81
interior, 81
intermediate value theorem, 115
interpolation lemma, 35
interval, 89
intuitionism, 71
inverse function, 117

Kronecker, Leopold, 21

Lebesgue integral, 16
Leibniz, Gottfried Wilhelm, 105
limit, 41, 59
limit point, 55
linear subspace, 89
Lipschitz-condition, 128
localization lemma, 37
located sequence, 65
located set, 71
Lorenzen, Paul, 70

maximum, 112
mediant, 1
metric, 49
metric space, 49
minimum, 112
mixture, 59
monotony, 115
multiplication, 22, 111

neighborhood, 80
nesting lemma, 35

open interval, 89
open set, 81
order, 22, 27
outer point, 81
overlapping lemma, 62

Pólya, Georg, 17, 18
Perron integral, 17

pndulum number, 25
Poincaré, Henri, 12
point, 49
pointwise continuity, 105
pointwise convergence, 118
positive-distance-theorem, 79
principle of permanence, 43
Pythagoras of Samos, 3

ray, 89
real number, 24
recursive constructivism, 70
Riemann integral, 15
root, 118
rounded approximation, 61
rounding, 23
rounding lemma, 61
ruler-scale-function, 96
Russell, Bertrand, 12

separable metric space, 72
set, 50
strict monotony, 115
subset, 50
subspace, 50
sufficient approximation, 63
supremum, 90
symmetry of the metric, 50

theorem about the infimum, 67, 72
theorem about the supremum and the infimum, 107
theorem of Arzelà and Ascoli, 124
theorem of Baire, 82
theorem of Bolzano, 108
theorem of Brouwer, 109
theorem of Cantor, 83
theorem of Cauchy, 56
theorem of Dini and Brouwer, 122
theorem of Eudoxos and Archimedes, 36
theorem of Heine and Borel, 77
theorem of Lebesgue, 16
theorem of Weierstrass, 90
theorem of Weyl and Brouwer, 102

totally bounded sequence, 64
totally bounded set, 71
triangle inequality, 33, 34, 49–51, 53

unbounded interval, 90
uniform continuity, 105
uniform convergence, 118

value, 95
van Dalen, Dirk, 13

Weyl, Hermann, 1, 12–15, 17, 18

Printed by Publishers' Graphics LLC